MW00465738

WARBLERS

and Other Songbirds of North America

WARBLERS

and Other Songbirds of North America

A Life-size Guide to Every Species

PAUL STERRY

Warblers and Other Songbirds of North America.

Published in 2017 by
Harper Design
An Imprint of HarperCollins*Publishers*
195 Broadway
New York, NY 10007
Tel: (212) 207-7000
Fax: (855) 746-6023
harperdesign@harpercollins.com
www.hc.com

Distributed throughout the world by
HarperCollins Publishers
195 Broadway
New York, NY 10007

ISBN 978-0-06-244681-7
Library of Congress Control Number 2016958608

Photograph page 1: Blue Jay
Photograph page 3: Eastern Bluebird
Photograph opposite: American Robin

Edited and designed by D & N Publishing, Wiltshire, UK
Printed in China

First Printing, 2017

Contents

The aim and scope of this book

Warblers and Songbirds of North America is a photographic field guide to the region's richly varied songbirds. The geographical area covered by the book extends from the Arctic in the north, south to the Mexican border. More than 250 species have been included in the book, this range covering the most regularly encountered songbirds in the USA and Canada.

The photographs used throughout the book have been chosen to show important identification features and to depict a bird's typical posture, be that perched or standing. As many plumage variations as possible have been included. For every species entry, at least one image shows the bird in question, or the head and bill in the case of large species, depicted life-size. The images are gloriously detailed, and the text that complements the photographs has been written as much with the beginner in mind as the experienced birdwatcher.

For each species, the main text contains descriptions of plumage and structural features that are useful for identification, plus further

A typical species description from the book.

information about habits and behavior. In addition, a fact file section covers key details for each species: common name; scientific name; length (an average, measured from bill tip to the end of the tail); food; habitat (or habitats, if these differ seasonally); status; and voice.

What makes birds special?

Apart from bats, birds are the only vertebrates capable of flight. The ancestors of modern birds took to the air some 150 million years ago, and since that time the ability to fly has allowed them to occupy almost every terrestrial habitat on Earth, and many aquatic ones too.

For birds, flight would not be possible without feathers, but these lightweight, tough and resilient structures are also vital for thermal insulation. In addition, contours, shapes, patterns and colors confer species and gender identity on their owners, and camouflage is also important for many species. Unsurprisingly, there are different feathers on a bird's body that fulfill a range of functions, those associated with flight being structurally different from those that insulate.

In common with their reptilian ancestors, birds lay eggs, inside which their young develop. Eggs are laid in a nest; in the case of songbirds, the location and structure of the nest can range from a rudimentary cup-shaped structure sited in a cavity on the ground or in a treehole, to an intricately woven basket suspended from foliage. The eggs themselves are protected by a hard, chalky outer shell and the developing chick gains its nutrition from the egg's yolk.

What is a songbird?

Songbirds are members of the order Passeriformes. This group of birds, commonly known as passerines, comprises a range of bird families; it is our most varied group of birds both in terms of numbers of species and diversity of appearance and habit preferences. Songbird members range in size from the tiny Golden-crowned Kinglet (our smallest bird) to the massive Common Raven.

Songbirds have feet that allow them to perch with ease. Three toes point forward and one faces back; this provides support and allows the bird in question to stand upright on level ground, and enables it to grasp comparatively slender twigs and branches with a sure grip.

As their name suggests, songbirds are—to a greater or lesser degree—extremely vocal, and males of some species are among the finest songsters in the bird world. Many territorial males advertise ownership of breeding grounds and attract and retain mates by loud and diagnostic songs. And all species have a repertoire of calls that serve a variety of behavioral functions, including alarm (for example, at the presence of a predator) or contact (with other members of the species in feeding flocks or on migration).

Songbird diet is as varied as the appearance of the birds themselves, but for many species small invertebrates are important for at least part of the year—typically the spring and summer months, when nesting is taking place. Some songbird families, such as warblers, feed almost exclusively on invertebrates, while sparrows and buntings rely to a great degree on seeds as a source of nutrition, particularly during the winter months. Many crow family members are arch scavengers that, to a certain extent, have predatory habits too. But in shrikes, the predilection for live prey reaches its apogee, the birds behaving like miniature raptors and even having hook-tipped bills to aid dismembering victims.

Most songbirds lead rather solitary lives during the breeding season and nest in relative isolation from pairs of the same species. However, outside the breeding season some form sizable flocks that migrate, feed, and roost together. There is some truth in the saying that there is safety in numbers, because there are plenty of eyes on the lookout for danger.

In some songbird species, visual differences between the sexes are subtle (to our eyes at least). Behavioral differences obviously play an important role in gender recognition for the birds themselves, but in the case of certain species it is only when a male is heard singing or when nesting behavior is observed that we as observers can be certain of the sex of a bird. However, among many songbird groups there are striking differences in plumage, although often these differences are more apparent in breeding plumage than during the winter months.

Bird topography

Birdwatchers give precise names to distinct parts of a bird's body, both to the bare parts (legs and bill, for example) and areas of feathering (wing coverts, primaries, and the like). These terms have been used throughout this book to ensure precision and to avoid ambiguity about what is being described or discussed. An understanding of this terminology helps the reader interpret the descriptive text in the book. It is also helps when talking about bird identification with other birdwatchers, and is useful in the process of identification in the field. The annotated photograph below shows the important anatomical and topographical features of a typical songbird, and a glossary of terms overleaf helps with the learning process.

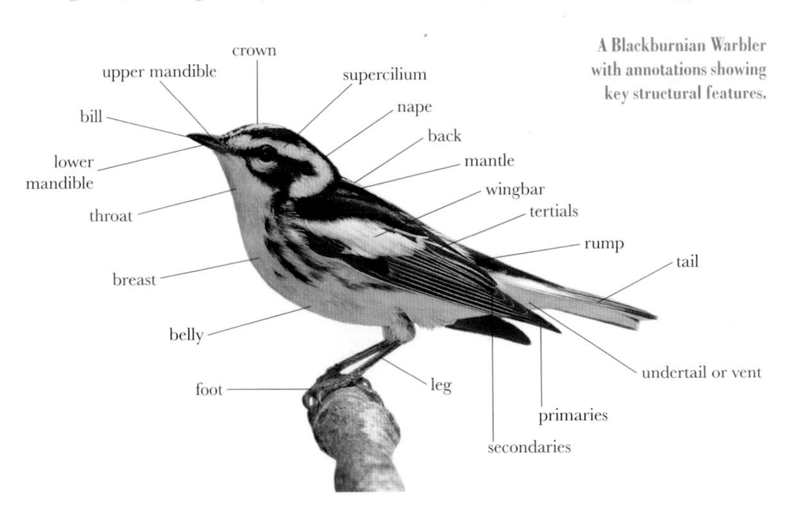

A Blackburnian Warbler with annotations showing key structural features.

Glossary

Adult A fully mature bird.
Bill The beak.
Coverts Areas of contour feathers found on the upperwing, underwing, uppertail, and undertail.
Eyering A ring of feathers, often colorful, that surrounds the eye.
First-winter A bird's plumage in its first winter after hatching.
Forewing The leading edge of the upperwing.
Immature A bird that is any age younger than an adult.
Juvenile A young bird with its first set of full feathers.
Lores The area between the eye and the bill.
Malar A band or stripe of feathers on the side of the throat, in front of and below the submustachial stripe.
Mandibles The two parts of a bird's bill: upper and lower.
Mantle Feathers covering the back.
Migrants Birds that have different, geographically separate breeding grounds and winter quarters.
Mustachial stripe A stripe that runs from the bill to below the eye, fancifully resembling a mustache.
Nape The hind neck.
Orbital ring Ring of bare skin around the eye, often brightly colored.
Primaries The main flight feathers found on the outer half of the wing.
Scapulars A group of feathers that form the "shoulder" of the bird between the back and folded wing.
Secondaries A group of relatively large flight feathers that form the inner part of the wing.
Species A taxonomic description relating to a population, members of which breed with one another but not with others. A species' scientific name is binomial, comprising the genus name first, followed by the specific name; taken together, the name is unique.
Submustachial stripe The contrasting line of feathers below the mustachial stripe.
Supercilium A stripe that runs above the eye.
Tertials The innermost flight feathers.
Vent The area underneath the tail, covered by the undertail coverts.
Wingbar A bar or band on the wings, created by aligned pale feather tips, often those of the wing coverts.

Olive-sided Flycatcher

Contopus cooperi

This species is a rather dull-looking flycatcher with a plump body and upright posture when perched. The bill is relatively large and dark, and the end of the tail is forked. In plumage terms, the sexes are similar, with dull olive-brown upperparts and a subtly darker tail and wings. It has faint pale wingbars, there is a subtle pale eyering, and note the white feather tufts on the side of the rump. The flanks are dark-streaked and on the underparts a broad whitish band extends from the throat down the center of the breast to the undertail. Overall, juveniles are a subtly warmer brown than adult birds.

The Olive-sided Flycatcher is a breeding species across northern temperate and western North America; it is present from May to August. It migrates south outside the breeding season and spends the rest of the year in South America. It specializes in catching flying insects on aerial forays. It often uses a dead branch as a lookout post.

adult life-size

FACT FILE

LENGTH 7.5 in (19 cm)

FOOD Insects

HABITAT Northern boreal forests and conifer forests farther south in range

STATUS Widespread and locally common summer visitor

VOICE Song is a rapid *quip-wee-ber*

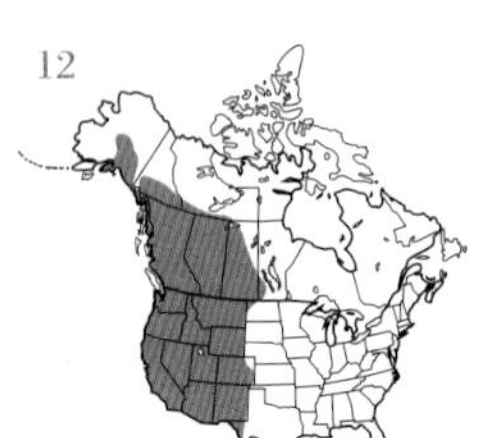

Western Wood-pewee

Contopus sordidulus

Compared to other flycatcher species, the Western Wood-pewee has relatively longer wings. Its legs are dark and the bill is mostly dark but with a dull yellow base to the lower mandible. The sexes are similar. Adults are gray-brown above with a blackish tail and wings. The wings show subtle pale wingbars and pale fringes to the inner flight feathers. The underparts are gray-brown, paler than the upperparts and palest on the throat, grading to whitish on the belly. Juveniles are similar to adults, but brighter looking and with buffish wingbars and fringes to the inner flight feathers.

The Western Wood-pewee is present as a breeding species across the western half of North America from May to September. It spends the rest of the year in northern South America. Birds typically perch at mid-level in open woodland. They sit motionless for long periods on lookout perches before flying out to catch an insect. Call and song offer the best chances of separation from very similar Eastern Wood-pewee in the field in regions where both might occur.

adult

adult life-size

FACT FILE

LENGTH 6.25 in (16 cm)

FOOD Insects and other invertebrates

HABITAT Open woodland

STATUS Widespread and locally common summer visitor

VOICE Song is a *tswee-tsee-tseet*. Call is a shrill, downslurred *psee-err*

Eastern Wood-pewee

Contopus virens

The legs are dark and the bill has a mostly dark upper mandible and dull orange flush to lower mandible. The sexes are similar. Adults are gray-brown above with a blackish tail and wings. The wings show subtle pale wingbars and pale fringes to the inner flight feathers. The underparts are gray-brown, paler than the upperparts and palest on the throat, grading to whitish on the belly. Juveniles are similar to adults, but brighter looking and with buffish wingbars and fringes to the inner flight feathers.

The Eastern Wood-pewee is present as a breeding species across the eastern half of North America from May to September. It spends the rest of the year in northern South America. Birds typically perch at mid-level in open woodland. They sit motionless for long periods on lookout perches before flying out to catch an insect.

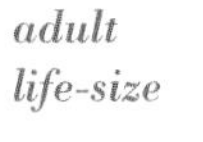

adult
life-size

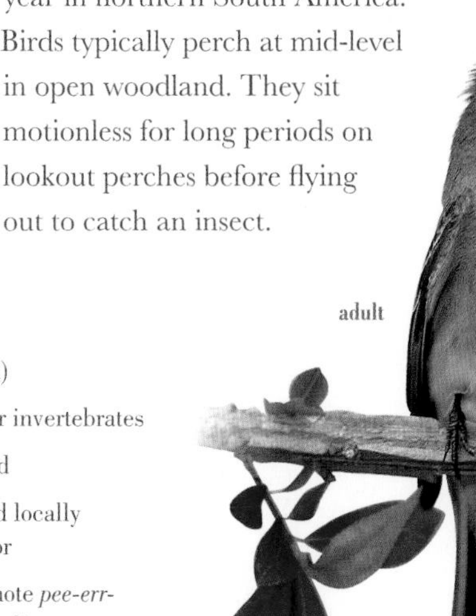

adult

FACT FILE

LENGTH 6.25 in (16 cm)

FOOD Insects and other invertebrates

HABITAT Open woodland

STATUS Widespread and locally common summer visitor

VOICE Song is a three-note *pee-err-wee*. Calls include whistling notes and a sharp *prrt*

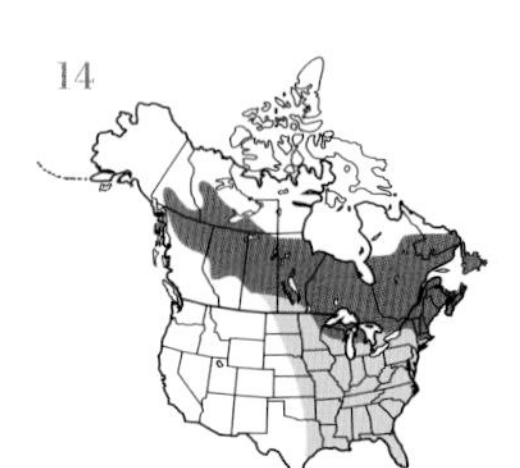

Yellow-bellied Flycatcher

Empidonax flaviventris

The Yellow-bellied Flycatcher adopts an upright stance when perched. Like related *Empidonax* species, it has a relatively large head and short tail. The sexes are similar. Adults have dull yellowish-green upperparts, and paler yellowish underparts with a darker band across the breast. The head has a pale eyering and the bill's lower mandible is dull pinkish orange. Compared to the rest of the plumage, the wings are dark, with two white wingbars and pale fringes to the inner flight feathers. Juveniles are similar to adults, although brighter overall and with yellowish, not white, wingbars.

The Yellow-bellied Flycatcher is present as a breeding species from June to August across northern latitudes of North America. It spends the rest of the year in Central and South America. In the breeding season it favors boreal forests, especially dense growths of spruce. Although not shy, it likes to perch in shade, making it easily overlooked; listen for its call and song to detect its presence.

adult

FACT FILE

LENGTH 5.5 in (14 cm)

FOOD Insects and other invertebrates

HABITAT Northern forests

STATUS Widespread and locally common summer visitor

VOICE Song is a shrill, two-note *tchWeek*. Call is a whistled *tchWwee*

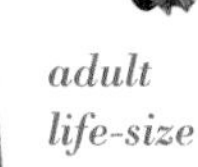

adult life-size

Acadian Flycatcher

Empidonax virescens

The Acadian Flycatcher is a rather plump-bodied flycatcher with a relatively large bill. The sexes are similar. Adults are dull yellow-green on the head, neck, and back, with paler—almost whitish—underparts that show a dull yellow-buff band across the breast. The wings are contrastingly dark, with two white wingbars. The head has a subtle whitish eyering and the lower mandible is pinkish orange. Juveniles are similar to adults but the wingbars are buffish, not white.

The Acadian Flycatcher is present as a breeding species in North America, mainly from April to August, its range being east and southeast U.S.A. It spends the rest of the year in Central and South America. In its favored forested habitat it usually perches mid-level, making flycatching forays for flying insects or picking invertebrates off the foliage. The distinctive song and call are often the first clues to the species' presence in an area of forest.

adult
life-size

FACT FILE

LENGTH 5.5 in (14 cm)

FOOD Insects and other invertebrates

HABITAT Deciduous forests, especially near water

STATUS Widespread and locally common summer visitor

VOICE Song is a loud *pe-Pswerp*. Call is a sharp *pweep*

Alder Flycatcher

Empidonax alnorum

adult

The Alder Flycatcher is very similar to the Willow Flycatcher (p.17), with their songs and calls offering the best chance of certain identification; silent birds are often not separable. The sexes are similar. Adults have dull yellow-green upperparts, and pale underparts with a gray-green flush across the chest. There is a white eyering and the bill has a pinkish-orange lower mandible. The wings are dark overall but with pale edges to the inner flight feathers and two white wingbars. Juveniles are similar to adults but the wingbars are buff, not white.

The Alder Flycatcher is present as a breeding species across northern latitudes of North America from June to August. It spends the rest of the year in South America. It favors damp woodland thickets where species of alders, willows, and birches flourish, and usually perches near the tops of trees, watching for passing insects. Although the species is relatively easy to observe, separating it from the Willow Flycatcher is more of a challenge.

adult life-size

FACT FILE

LENGTH 5.75 in (14.5 cm)

FOOD Insects and other invertebrates

HABITAT Deciduous woodland

STATUS Widespread and locally common summer visitor

VOICE Song is a harsh, repeated *rrre-BEE-ah*. Call is a sharp *piip*

Willow Flycatcher

Empidonax traillii

The Willow Flycatcher closely resembles the Alder Flycatcher (p.16). The sexes are similar. Adults from the eastern half of the species' range have olive-gray upperparts (subtly greener than in Alder Flycatcher); birds breeding in the west have browner and darker upperparts. All adults have pale underparts with a soft gray-green wash across the breast. There is a pale eyering and the bill has a pinkish-orange lower mandible. The wings are dark overall with pale fringes to the inner flight feathers and two white wingbars. Juveniles are similar to adults but the wingbars are buff, not white.

The Willow Flycatcher is present as a breeding species across central and western North America from June to August. It spends the rest of the year in Central and South America. It favors damp woodland where willows flourish but alders and birches may also be present. It usually perches near the tops of trees, watching for passing insects, and sometimes hovers, gleaning insects from leaves. Silent birds can be hard to separate from the Alder Flycatcher.

adult
life-size

FACT FILE

LENGTH 5.75 in (14.5 cm)

FOOD Insects and other invertebrates

HABITAT Deciduous woodland

STATUS Widespread and locally common summer visitor

VOICE Song is a harsh, buzzing *fzz-byew*. Call is a sharp *whuit*

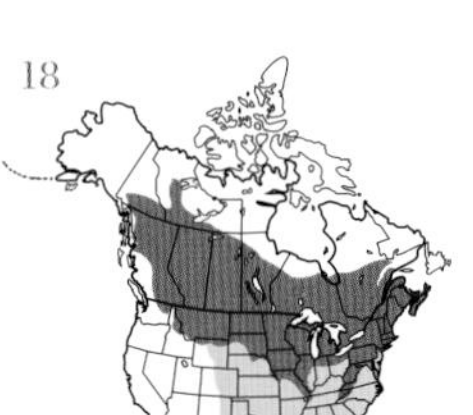

Least Flycatcher

Empidonax minimus

The Least Flycatcher is a compact, short-winged species with a relatively large head. The sexes are similar. Adults have dull olive-gray upperparts; the underparts are whitish, palest on the throat and belly, and with a gray band across the breast and a dull yellow-buff flush to the lower flanks. The dark wings have pale fringes to the inner flight feathers and two white wingbars. Juveniles are similar to adults but the wingbars are buff, not white.

The Least Flycatcher is present as a breeding species from May to August across much of the northern half of North America. It spends the rest of the year in Central America. It usually sits at mid-level in the tree canopy and often adopts an upright stance, regularly flicking its tail and wings. Insects are caught in flight; Least Flycatchers also hover and pick prey from the surface of leaves.

adult

adult
life-size

FACT FILE

LENGTH 5.25 in (13.5 cm)

FOOD Insects and other invertebrates

HABITAT Open deciduous woodland

STATUS Widespread and locally common summer visitor

VOICE Song is a repeated series of *che-bik, che-bik* phrases. Call is a sharp *whit*

Hammond's Flycatcher

Empidonax hammondii

Hammond's Flycatcher has relatively long wings and a proportionately large head and eyes. The sexes are similar. Adults have a gray head and neck, grading to gray-green on the back. The underparts are pale gray, palest on the throat; the pale belly has a faint yellow flush. The wings are dark overall with two pale wingbars. There is a white eyering, boldest behind the eye, and the bill is mainly dark but with a subtle orange base to the lower mandible. Juveniles are similar to adults but the wingbars are buff, not white, and the color at the base of the lower mandible is often bolder.

Hammond's Flycatcher is present as a breeding species from April to August, its range being western North America. It spends the rest of the year mainly in Central America. The species is very active and on the move much of the time, cocking its tail and flicking its wings as it goes.

adult

adult life-size

FACT FILE

LENGTH 5.5 in (14 cm)

FOOD Insects and other invertebrates

HABITAT Open conifer forests, often in mountains

STATUS Locally common summer visitor

VOICE Song is a disyllabic *tch-wip*. Call is a sharp *whit*

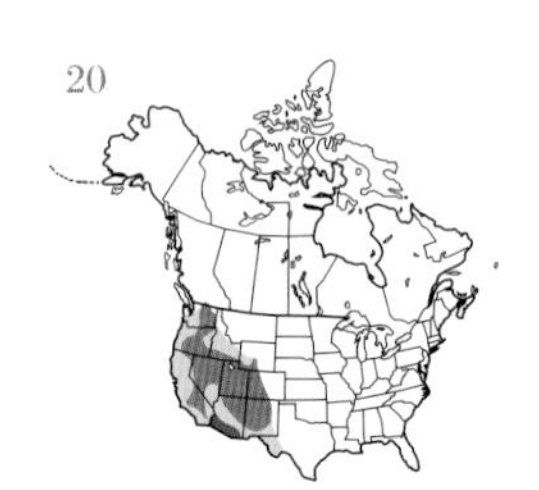

Gray Flycatcher

Empidonax wrightii

adult

The Gray Flycatcher has rather short wings and a relatively large head and eyes. The sexes are similar. Adults have a gray head, neck, and back; the underparts are pale gray, palest on the throat and belly. The head has a pale eyering, and the base of the bill's lower mandible is dull orange-pink with a contrasting dark tip. The wings are dark overall but with two bold pale wingbars. Juveniles are similar to adults but overall the plumage has a buffish wash.

The Gray Flycatcher is present as a breeding species in western interior North America from April to August. It spends the rest of the year mainly in Mexico. It usually adopts an upright posture when perched, pumping its tail up and down. It undertakes flycatching forays and also picks insects from foliage while hovering.

FACT FILE

LENGTH 6 in (15 cm)

FOOD Insects and other invertebrates

HABITAT Pinyon pine (*Pinus* spp.) woodland and sagebrush (*Artemisia* spp.)

STATUS Locally common summer visitor

VOICE Song is a repeated *chi-wip*. Call is a sharp *whit*

adult life-size

Dusky Flycatcher

Empidonax oberholseri

The Dusky Flycatcher is very similar to both Gray (p.20) and Hammond's flycatchers (p.19). Subtle behavioral as well as structural differences aid identification: It flicks its wings and tail (but does not pump the latter like Gray), and it is less active than Hammond's. Of the species trio, it has the shortest wings and the primaries do not extend beyond the rump when perched. The sexes are similar. Adults have olive-gray upperparts and pale gray underparts, palest on the throat and belly. There is a pale eyering and the base of the lower mandible is dull orange-pink. The wings are dark with two whitish wingbars. Juveniles are similar to, but brighter than, fall adults and have buffish, not white, wingbars.

The Dusky Flycatcher is present as a breeding species in western interior North America, mainly from May to August. It spends the rest of the year mainly in Mexico, with small numbers in southern Arizona. It usually adopts an upright posture when perched, and undertakes flycatching forays; it also hovers and picks insects from foliage.

FACT FILE

LENGTH 5.75 in (14.5 cm)

FOOD Insects and other invertebrates

HABITAT Mountain conifer forests

STATUS Locally common summer visitor

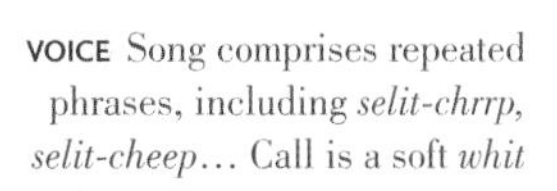

VOICE Song comprises repeated phrases, including *selit-chrrp, selit-cheep*... Call is a soft *whit*

adult life-size

Pacific-slope Flycatcher

Empidonax difficilis

The Pacific-slope Flycatcher is very similar to the Cordilleran Flycatcher (p.23) and the two are only really separable by noting the call and breeding range. The sexes are similar. Adults have olive-brown upperparts and pale underparts, palest on the throat and belly. Like Cordilleran, the relatively large eye is accentuated by a white eyering that broadens behind it. The broad-based bill has an orange-pink lower mandible. The wings are dark overall with pale fringes to the inner flight feathers and two pale wingbars. Juveniles are similar to adults but the wing feather fringes and wingbars are buff.

adult

The Pacific-slope Flycatcher is present as a breeding species mainly from May to August. Its breeding range is western North America and, as its name suggests, it is restricted to the coastal belt. It spends the rest of the year in Central America. It often adopts an upright posture when perched and engages in flycatching forays after flying insects.

adult life-size

FACT FILE

LENGTH 5.5 in (14 cm)

FOOD Insects and other invertebrates

HABITAT Damp conifer woodlands

STATUS Locally common summer visitor

VOICE Song usually comprises three repeated phrases: a thin *tsi*; a loud *tsche-wee*; and thin, sharp *pik*. Call is a slurred *tsweep*

Cordilleran Flycatcher

Empidonax occidentalis

This species is almost identical to the Pacific-slope Flycatcher (p.22), and silent birds—particularly on migration—can be impossible to identify with certainty. The calls and breeding range offer the best clues for separation. The sexes are similar. Adults have olive-brown upperparts (subtly browner than Pacific-slope) and pale underparts, palest on the throat and belly. The eye is accentuated by a white eyering that broadens behind it, and the bill is marginally longer than that of Pacific-slope. The wings are dark, with pale fringes to the inner flight feathers and two dull white wingbars. Juveniles are similar to adults but the wing feather fringes and wingbars are buff.

The Cordilleran Flycatcher is present as a breeding species mainly from May to August; its interior upland breeding range does not overlap that of the Pacific-slope Flycatcher. It spends the rest of the year in Central America. The species adopts an upright posture when perched.

adult

FACT FILE

LENGTH 5.5 in (14 cm)

FOOD Insects and other invertebrates

HABITAT Conifer forests in the Rocky Mountains

STATUS Locally common summer visitor

VOICE Song typically comprises a thin *tsee*, a chirping *see-oo*, and a sharp *pik*. Call is a thin, whistled, disyllabic *tsee-seet*

adult
life-size

Black Phoebe

Sayornis nigricans

The Black Phoebe is a distinctive, easily recognizable flycatcher, being dumpy-bodied and essentially black and white. The sexes are similar. Adults have a black head and chest, grading to dark gray on the back. The wings are black overall but with a hint of pale on the wingbars and subtly pale fringes to the inner flight feathers. The rather long tail is mainly black but with white outer feathers. The belly and undertail are white. Juveniles are similar to adults but the wing coverts and back feathers have brown fringes.

The Black Phoebe is resident throughout the year in southwest U.S.A. It is easy to observe well because typically it is almost oblivious to observers. It usually perches in the open and flies out to catch passing insects. A characteristic behavior is that it wags its tail up and down when perched.

adult life-size

adult

FACT FILE

LENGTH 6.75 in (17 cm)

FOOD Insects and other invertebrates

HABITAT Open woodland, parks, and gardens, usually near water

STATUS Locally common resident

VOICE Song is a repeated two-phrase *tch-wee, tch-sew*. Call is a sharp *tsiip*

Eastern Phoebe

Sayornis phoebe

The Eastern Phoebe is a plump-bodied flycatcher whose plumage lacks any really distinctive features. The sexes are similar. Adults are mainly gray-brown above, darkest on the head. The wings are blackish with two whitish wingbars and pale fringes to the inner flight feathers. The underparts are dull whitish but with a gray wash to the flanks; freshly molted birds in fall show a variable yellow-buff wash to the belly. Juveniles are similar to adults but with buff wingbars and a more obvious yellow wash on the belly.

The Eastern Phoebe is a breeding species across much of eastern North America mainly from April to September. It spends the rest of the year in southeast U.S.A. and Mexico. Perched birds often pump their tail downwards, with a swaying motion. The species flycatches in flight, usually from the vantage point of a low perch. Eastern Phoebes often favor manmade habitats, and nest under bridges or on buildings.

adult life-size

adult

FACT FILE

LENGTH 6.75 in (17 cm)

FOOD Insects and other invertebrates

HABITAT Woodland, parks, and gardens

STATUS Widespread and common summer visitor

VOICE Song is a whistled two-phrase *sphee-dip, sphee-werr*. Call is a sharp *chip*

Say's Phoebe

Sayornis saya

Say's Phoebe has a relatively large, long tail. The sexes are similar. Adults have mainly gray-brown upperparts that are darkest on the head. On the underparts the throat and breast are gray-buff, and this color grades to a dull peachy orange on the belly and undertail. The wings are dark overall but with subtly pale wingbars and pale fringes to the inner flight feathers. In flight, note that the underwing coverts are flushed peachy orange. Juveniles are similar to adults but the plumage is browner overall and the wing feather fringes are orange-buff.

adult

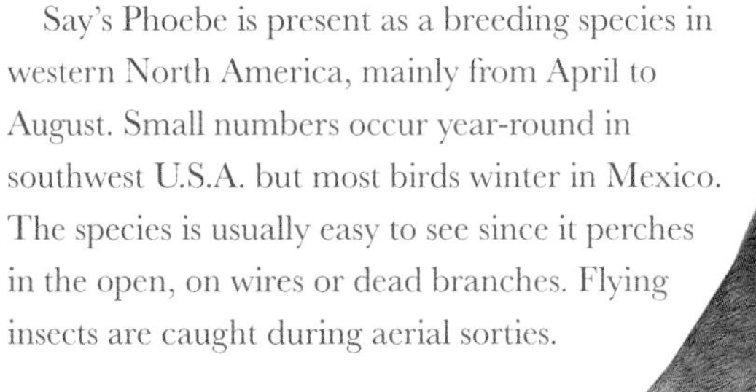

Say's Phoebe is present as a breeding species in western North America, mainly from April to August. Small numbers occur year-round in southwest U.S.A. but most birds winter in Mexico. The species is usually easy to see since it perches in the open, on wires or dead branches. Flying insects are caught during aerial sorties.

adult life-size

FACT FILE

LENGTH 7.5 in (19 cm)

FOOD Insects and other invertebrates, but also berries

HABITAT Wide range of open habitats, including deserts, tundra, and grassland

STATUS Widespread and common summer visitor

VOICE Song is a whistled and repeated *pit-seer.* Call is a sharp *pe-eer*

Vermilion Flycatcher

Pyrocephalus rubinus

The Vermilion Flycatcher is a distinctive bird. The sexes are dissimilar and males are unmistakable. Adult males have a bright red throat, underparts, and crown, with a dark brown mask, nape, back, wings, and tail. Adult females have gray-brown upperparts, a pale supercilium, dark wings with two pale wingbars, and a dark tail. The underparts are pale overall, with dark streaking on the breast and flanks, and a pinkish-orange flush to the belly and undertail. Juveniles are similar to adult females but have spots, not streaks, on the breast and flanks, and lack color on the underparts. Immature birds in their first year show plumage characteristics intermediate between juvenile and adult plumages.

female

The Vermilion Flycatcher is present as a summer visitor to the southern states of the U.S.A., mainly from May to September. A few birds are present year-round but most winter in Central America. This colorful species usually perches conspicuously and is indifferent to people, making observation both easy and rewarding.

male
life-size

FACT FILE

LENGTH 6 in (15 cm)

FOOD Insects and other invertebrates

HABITAT Open woodland and parks, usually near water

STATUS Summer visitor, locally common within its restricted North American range

VOICE Song is a series of sharp *pit-a-see* notes, ending in a trill, sometimes given in flight. Call is a thin *psee*

Ash-throated Flycatcher

Myiarchus cinerascens

The Ash-throated Flycatcher has a relatively large head and a slim-bodied appearance that is accentuated by its long neck and tail. The sexes are similar. Adults have a brown cap, separated from the gray-brown back by a paler nape. The underparts are very pale overall and flushed lemon yellow, except for the pale gray throat and chest. The wings are blackish overall but with two pale wingbars, pale fringes to the secondaries, and rufous-tinged primaries. The tail feathers are mostly rufous but, from below, the feather tips are dark gray. The bill and legs are dark. Juveniles are similar to adults but less colorful, and the secondary feather fringes are rufous, not pale.

adult
life-size

FACT FILE

LENGTH 7.5–8.5 in (19–21.5 cm)

FOOD Insects, other invertebrates, and fruit

HABITAT Deserts and dry woodland

STATUS Locally common summer visitor

VOICE Song (sung at dawn) is a series of *Ke-reer* notes, similar to call. Call is a subdued *ke-Brik*

adult

adult

The Ash-throated Flycatcher is present as a breeding species in southwest U.S.A., mainly from April to August. Although a few remain in the region year-round, most are found in Central America at other times of the year. The species sits upright when perched, and its feeding habits include catching passing insects in flight and gleaning prey and fruit from foliage. It also scans the ground for prey or fallen fruit, onto which it drops.

Great Crested Flycatcher

Myiarchus crinitus

The Great Crested Flycatcher is a relatively large, slim-bodied species. The sexes are similar. Adults have a dark brown hood, nape, and back. The underparts comprise a dark gray face, throat, and breast, with a clear separation from the bright yellowish belly and undertail. The wings are dark overall but show two pale wingbars and rufous-tinged primaries. The tail is mainly rufous, both above and below. The legs are black and the bill is black overall but with a dark brown base. Juveniles are similar to adults but with subdued colors overall, and rufous wingbars and flight feather edges.

adult
life-size

FACT FILE

LENGTH 7–8.5 in (18–21.5 cm)

FOOD Insects, other invertebrates, and fruit

HABITAT Wide range of wooded habitats

STATUS Common summer visitor

VOICE Song (sung at dawn) comprises a series of *whu-eep* call-like notes. Calls include an upslurred *whu-eep* and a harsh *chrrrt*

adult

adult

The Great Crested Flycatcher is present as a breeding species across most of temperate eastern North America from April to September. It spends the rest of the year in Central and South America. It perches in an upright posture and feeds by flycatching passing insects, gleaning insects from foliage, and dropping to the ground for prey.

Brown-crested Flycatcher

Myiarchus tyrannulus

The Brown-crested Flycatcher could be confused with the Great Crested Flycatcher (p.30), although their breeding ranges and preferred habitats do not overlap. Furthermore, Brown-crested is a subtly larger bird with a larger, mostly black bill. The sexes are similar. Adults have a gray-brown hood, nape, and back. The face, throat, and breast are pale gray, and the underparts are otherwise pale yellow. The wings are blackish overall but with two dull wingbars, rufous fringes to the primaries, and pale fringes to the other flight feathers. The tail is mainly rufous, both above and below. Juveniles are similar to adults but the wingbars and most flight feather fringes are rufous.

adult
white

The Brown-crested Flycatcher is present as a breeding species mainly from May to August, within a limited range in the southernmost states of the U.S.A. It spends the rest of the year in Mexico. It adopts an upright posture when perched, and feeds by flycatching and dropping onto prey on the ground.

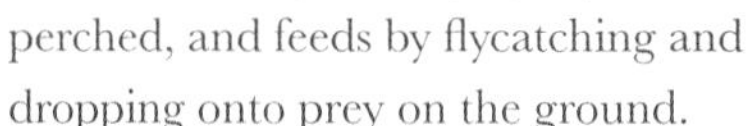

adult

FACT FILE

LENGTH 7.5–9 in (19–23 cm)

FOOD Insects, other invertebrates, and fruit

HABITAT Deserts, with plenty of scrub and cacti

STATUS Locally common summer visitor

VOICE Song (sung at dawn) comprises a series of call-like notes plus a variety of whistles. Call is a liquid *wrrt* or *wrrt-ado*

Tropical Kingbird

Tyrannus melancholicus

Compared to its close relatives in the genus *Tyrannus*, the Tropical Kingbird is paler overall and has a larger bill, shorter wings, and a forked tail. The sexes are similar. Adults have a gray head, whitish throat, and dark mask. The back is greenish grey and the breast is greenish yellow, grading to bright yellow on the belly and undertail. The wings are brown with pale buff feather margins. In flight, the bright yellow underwing coverts are striking. Juveniles are similar to adults but the wing feather margins are more obviously buff.

As its name suggests, the Tropical Kingbird's main range is south of the region covered by this book. It is present in southern Arizona as a breeding species mainly from April to August; in the Lower Rio Grande Valley of Texas it is present year-round. The species perches on bare branches and roadside wires, from where it flycatches and drops to the ground after prey.

adult

FACT FILE

LENGTH 9–9.5 in (23–24 cm)

FOOD Insects, other invertebrates, and fruit

HABITAT Open woodland

STATUS Local summer visitor and rare resident

VOICE Song (sung at dawn) comprises a series of *tlip* notes, ending with a call-like trill. Call is a chirruping trill

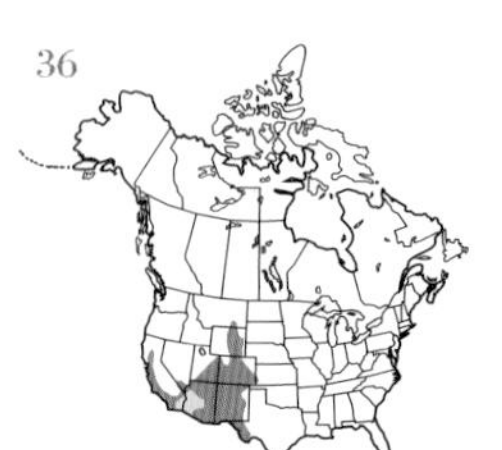

Cassin's Kingbird

Tyrannus vociferans

Cassin's Kingbird is overall darker than the similar Western Kingbird, with different vocalizations and habitat preferences. The sexes are similar. Adults have a dark gray head, back, and chest, and a bold white throat patch offset by the blackish lores. An orange-red crown patch is invariably hidden. The underparts are bright yellow and separated distinctly from the gray breast. The brown wings have pale feather margins and the tail is dark brown but with a subtle pale tip. Juveniles are similar to adults but the wing feathers have buff margins and the tail's pale tip is indistinct.

Cassin's Kingbird is present as a breeding species mainly from April to August. The majority of the population spends the winter months in western Mexico, although small numbers linger in southern California. The species is usually found at higher elevations—typically mountainside woodland—than Western Kingbird, a similar species with which its general range overlaps.

adult

FACT FILE

LENGTH 8.5–9 in (21.5–23 cm)

FOOD Insects, other invertebrates, and fruit

HABITAT Upland oak and pine woodland

STATUS Locally common summer visitor

VOICE Song (sung at dawn) is a series of *chrr-chrr-chrr…* notes. Call is a sharp *cha-Beer*

adult
life-size

Western Kingbird

Tyrannus verticalis

The Western Kingbird appears rather pale overall when perched. Seen from above in flight, the contrast between the rather pale body and dark wings and tail is striking. The sexes are similar. Adults have a mostly pale gray head with a subtle dark "mask" through the eye and a whitish cheek. An orange central crown patch is invariably hidden. The back is pale greenish gray, the chest is pale gray, and the underparts—including the underwing coverts—are pale lemon yellow. The dark wings have pale feather margins, and the dark tail has pale outer feather margins. Juveniles are similar to, but paler than, adults.

The Western Kingbird is present as a breeding species mainly from April to August. It spends the rest of the year in Central America. It is the most widespread kingbird in western North America and is easy to observe since it often perches on wires or dead branches beside farmland roads. Its flycatching sorties attract attention, as does the male's tumbling aerial courtship display.

FACT FILE

LENGTH 8–9.5 in (20–24 cm)

FOOD Insects, other invertebrates, and fruit

HABITAT Lightly wooded open country and farmland

STATUS Widespread and common summer visitor

VOICE Song (sung at dawn) comprises a series of call-like *chip* notes. Calls include a sharp *chip* and a trilling chatter

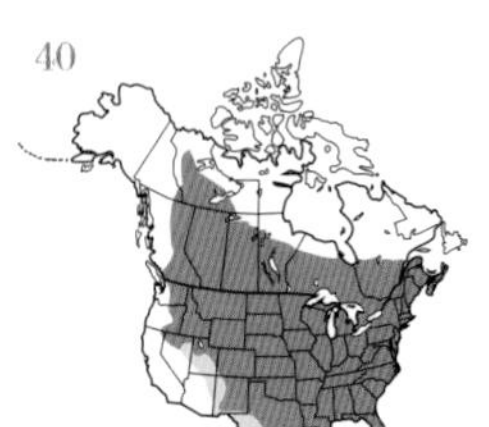

Eastern Kingbird

Tyrannus tyrannus

The Eastern Kingbird is a familiar black and white songbird. The sexes are similar. Adults are essentially black above and white below. The black hood, neatly defined by the white throat, grades to dark blue-gray on the back and wings; the wing feathers have whitish margins. A reddish-orange crown stripe is invariably hidden. The black tail has a white terminal band. Although the underparts are mostly white, a diffuse pale gray chest band is present. The bill and legs are dark. Juveniles are similar to adults, although the upperparts are subtly tinged brown.

adult life-size

The Eastern Kingbird is present as a breeding species mainly from May to August across most of temperate North America except the far west. It spends the rest of the year in South America. Great views can usually be obtained because it is often indifferent to people and perches on roadside wires and fences. It catches flying insects by making aerial sorties.

adult

adult

FACT FILE

LENGTH 8–9 in (20–23 cm)

FOOD Insects and other invertebrates

HABITAT Wide range of open habitats

STATUS Widespread and common summer visitor

VOICE Song (sung at dawn) comprises a much-repeated series of call-like trilling phrases. Calls include various trills and a metallic, rasping *kedzee-kedzee*…

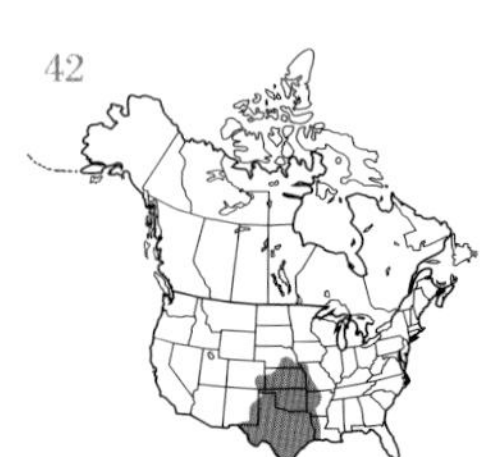

Scissor-tailed Flycatcher

Tyrannus forficatus

With its strikingly long tail and aerobatic habits, the Scissor-tailed Flycatcher is unmistakable. The sexes are similar, but an adult male's tail streamers are longer than those of the female. Adults have a pale blue-gray head, back, and breast, black wings with white feather margins, and a deeply forked tail with streamer-like outer feathers. The underparts are pale, flushed pinkish orange on the belly, undertail, and underwing coverts. In flight, the deep red axillaries ("armpits") can be seen. Juveniles are much paler than adults and the tail's outer feathers are short, not streamer-like.

The Scissor-tailed Flycatcher is present as a breeding species mainly from April to August, with Texas and Louisiana being at the center of its summer range. It spends the rest of the year in Central America. It often perches on overhead wires and is usually indifferent to people, making it easy to see. The species catches flying insects, typically during aerial sorties. The male performs his spectacular courtship display in flight.

adult

FACT FILE

LENGTH 10–15 in (25.5–38 cm)

FOOD Insects and other invertebrates

HABITAT Farmland and open country

STATUS Locally common summer visitor

VOICE Song (sung at dawn) comprises a series of call-like phrases. Calls include a sharp *wip* and a trilling *wip-kprrr*

Loggerhead Shrike

Lanius ludovicianus

The Loggerhead Shrike is a boldly marked predatory songbird. The sexes are similar. Adults have rich gray upperparts, the mantle having a white border. A dark "mask" extends through the eye and around the forehead, above the bill; it is defined above by a white margin. The underparts are pale overall, darkest on the breast and palest on the throat. The mainly black wings have a small white patch at the base of the primaries (most obvious in flight). The long tail is tapered and mainly black with white feather tips. Juveniles are similar to adults but the upperparts and underparts appear faintly barred or scaly.

FACT FILE

LENGTH 9 in (23 cm)

FOOD Insects, small mammals, and birds

HABITAT Farmland and open country with bushes and wires

STATUS Widespread but never common. Northern populations migrate south; southern birds are resident.

VOICE Song is a series of harsh, repeated chirping phrases. Call is a harsh *chaak*

adult

The Loggerhead Shrike occurs year-round in the south of its range, while northern birds are present in their breeding range mainly from May to September; they move south in fall. It is the smaller counterpart of the Northern Shrike (p.46), which is entirely migratory. Despite its size, the Loggerhead Shrike is a fearsome predator of large insects and small vertebrates; larger prey is often impaled on thorns or barbed wire and then dismembered. The species commonly perches on bushes and wire fences.

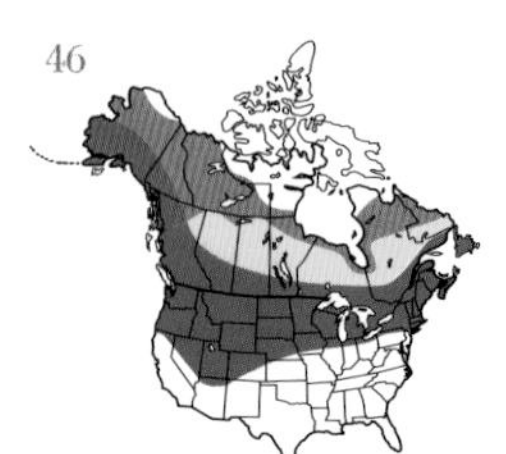

Northern Shrike

Lanius excubitor

The Northern Shrike is a larger, more heavily built cousin of the Loggerhead Shrike (p.44). The sexes are similar. Adults have pale gray upperparts and a black "mask" that reaches the bill but does not continue around the forehead (cf. Loggerhead). The underparts are very pale gray. The wings are mainly black but with a white patch at the base of the primaries; this is obvious in flight and more extensive than in Loggerhead. The long, tapered tail is mainly black but with white feather tips. Juveniles are similar to adults but the gray and white plumage elements are replaced by reddish buff; the underparts are barred faintly and the "mask" is faint. This plumage is usually largely replaced by adult-like plumage by early winter.

Northern Shrikes are present in their Arctic and boreal breeding range mainly from May to September. The species moves south to central North America for the rest of the year. This bold predator often perches on an overhead wire or dead branch. Larger prey items (small birds and mammals) are sometimes impaled on thorns or barbed wire before being dismembered.

adult

adult

adult life-size

FACT FILE

LENGTH 10 in (25.5 cm)

FOOD Insects, small mammals, and birds

HABITAT Tundra and taiga forest in summer; open country with scattered trees in winter

STATUS Widespread but never common

VOICE Song is a series of harsh squawks, chatters, and trills. Calls include a shrill *kreek, kreek*…

White-eyed Vireo

Vireo griseus

adult

The White-eyed Vireo is a well-marked but unobtrusive species. The sexes are similar. Adults have a greenish cap, back, and rump, and gray sides to the face and neck. The eye has a pale iris; it is emphasized by a yellow "spectacle" circling the eye and extending forward as a pale line, with a dark line below it. The underparts are pale, whitest on the throat and with a gray wash to the chest and a yellow wash on the flanks. The dark wings have two white wingbars. The bill is dark and stout, and the legs are blue-gray. Juveniles are similar to adults but paler, and with a dark iris and white "spectacle." Adult-like plumage and a pale iris are acquired in winter.

The White-eyed Vireo is present as a breeding species in eastern North America, mainly from April to September. The winter range extends from southeast U.S.A. to Mexico. The species is borderline secretive and usually easier to hear than to see. Learn its distinctive song to be certain of its presence in an area, then wait quietly and patiently for it to put in an appearance. Feeding birds forage in a deliberate manner.

adult life-size

FACT FILE

LENGTH 5 in (12.5 cm)

FOOD Insects and other invertebrates

HABITAT Dense deciduous woodland

STATUS Widespread and common summer visitor

VOICE Song is a series of loud phrases such as *chic, chip-ee-err-cheeo-chic.* Call is a harsh *shrrr*

Bell's Vireo

Vireo bellii

Bell's Vireo is a compact songbird that is rather warbler-like in its general appearance. Overall, western birds are paler and grayer than their eastern counterparts. Given this regional variation, however, the sexes are similar. Adults have overall greenish or greenish-gray upperparts, and pale underparts suffused with buffish yellow. The wings are subtly darker than the rest of the body, with a distinct pale wingbar and a more subtle one above it. The dark eye is emphasized by an incomplete white surround that extends forwards as a pale line to the base of the bill. Juveniles are similar to adults.

Bell's Vireo is present as a breeding species across the Midwest and southernmost southwest U.S.A. mainly from April to August. It spends the rest of the year in Mexico. The species is rather secretive and its presence is usually detected first by hearing its distinctive song. It feeds by foraging for insects and other invertebrates in dense foliage in a deliberate manner.

adult life-size

FACT FILE

LENGTH 4.75 in (12 cm)

FOOD Insects and other invertebrates

HABITAT Dense riverside scrub and woodland

STATUS Local and rather scarce summer visitor

VOICE Song is a rapid, chattering *chew-dlee, chew-dler*. Call is a thin *chee*

adult

Gray Vireo

Vireo vicinior

The Gray Vireo lacks any really distinctive plumage features and perhaps could be mistaken for a bulky warbler of some kind. However, within its breeding range there are few species with which it could be confused, so its drab appearance is actually a clue to its identity. The sexes are similar. Adults have blue-gray upperparts, darkest on the back, and whitish underparts, palest on the throat and flanks. The dark eye is accentuated by a faint pale eyering. The wings have faint pale margins to the flight feathers and a subtle pale wingbar. Juveniles are similar to adults.

The Gray Vireo is present as a breeding species in southwest U.S.A. mainly from May to August. It spends the rest of the year mainly in Mexico. Its presence is easiest to detect in an area by listening for its distinctive song. Although it is generally unobtrusive, the species feeds actively and flicks its tail from side to side, sometimes attracting the attention of observers as it does so.

adult

adult life-size

FACT FILE

LENGTH 5.5 in (14 cm)

FOOD Insects and other invertebrates

HABITAT Chaparral and arid scrub

STATUS Common summer visitor

VOICE Song is a repeated series of *tche-Woo, tche-Wee* phrases. Call is a rasping *chrrr*

Yellow-throated Vireo

Vireo flavifrons

adult

The Yellow-throated Vireo is a colorful and well-marked songbird with a stout bill and relatively large head. The sexes are similar, and juveniles resemble adults. Adults have yellowish-green upperparts, and are bright yellow on the lores, eyering, throat, and breast; the underparts are otherwise white. The dark wings have two white wingbars and pale margins to the inner flight feathers. The legs are blue-gray.

The Yellow-throated Vireo is present as a breeding species across much of eastern North America, mainly from April to August. It spends the rest of the year in Central America. Despite its colorful appearance, it can be hard to spot when feeding among foliage; often its presence is first detected by hearing its song or call. It searches for insects in a deliberate manner.

FACT FILE

LENGTH 5.5 in (14 cm)

FOOD Insects and other invertebrates

HABITAT Deciduous and mixed woodland

STATUS Widespread and common summer visitor

VOICE Song is a repeated *tchee-er-ee, tche-ett*. Calls include harsh *tcheh* notes

adult life-size

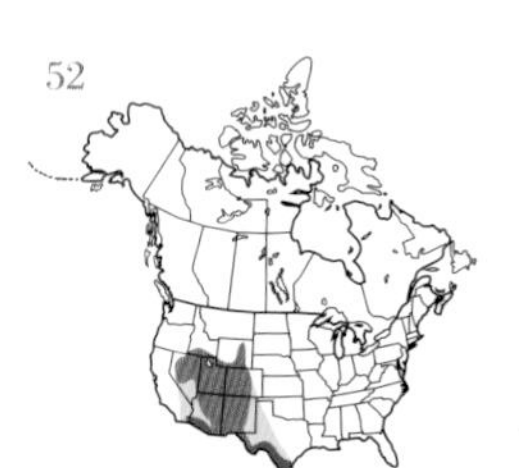

Plumbeous Vireo

Vireo plumbeus

adult

The Plumbeous Vireo is a chunky songbird with a relatively short but very stout bill; it is grayer overall than other vireo species. The sexes are similar. Adults have mainly dull gray upperparts, and whitish underparts that are washed gray on the flanks. The white "spectacles" around the eyes are striking features, and the dark wings have two white wingbars and pale margins to the inner flight feathers. Juveniles are similar to adults but with a subtle yellow suffusion on the flanks.

The Plumbeous Vireo is present as a breeding species in its well-defined Rocky Mountain range mainly from May to September. It spends the rest of the year mainly in Mexico. Given the rather open nature of its favored habitat, the species is relatively easy to see by the standards of other vireos. It hunts in a deliberate manner for insect prey.

FACT FILE

LENGTH 5.5 in (14 cm)

FOOD Insects and other invertebrates

HABITAT Forests in the Rocky Mountains

STATUS Locally common summer visitor

VOICE Song comprises a whistled *tchee-oee* followed by a churring *tchewee*. Call is a rasping *tche*

adult life-size

Cassin's Vireo

Vireo cassinii

adult

This species is very similar to both Blue-headed (p.54) and Plumbeous (p.52) vireos; its breeding range and habitat preferences help with separation. The sexes are broadly similar, although males are usually more brightly marked than females. All adults have a gray-green hood and a greenish-gray back. The underparts, including the throat, are mostly white with a dull yellow suffusion to the flanks. The white "spectacles" around the eyes are striking, and the dark wings have two pale (often subtly yellow) wingbars and pale margins to the inner flight feathers. The legs are blue-gray. Juveniles are similar to adults but overall duller.

Cassin's Vireo is present as a breeding species in North America mainly from May to August, its range restricted to western mountain ranges. At other times of the year it is found mainly in Mexico, with a limited presence in southern Arizona. Migrant birds seen away from their breeding range can be hard to identify with certainty.

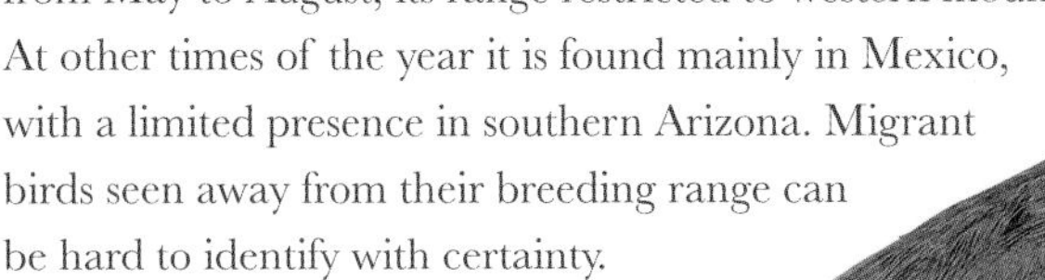

adult
life-size

FACT FILE

LENGTH 5 in (12.5 cm)

FOOD Insects and other invertebrates

HABITAT Open mountainside conifer forests

STATUS Common summer visitor

VOICE Song comprises a whistled *tchee-oee* followed by a churring *tchewee.* Call is a rasping *tche*

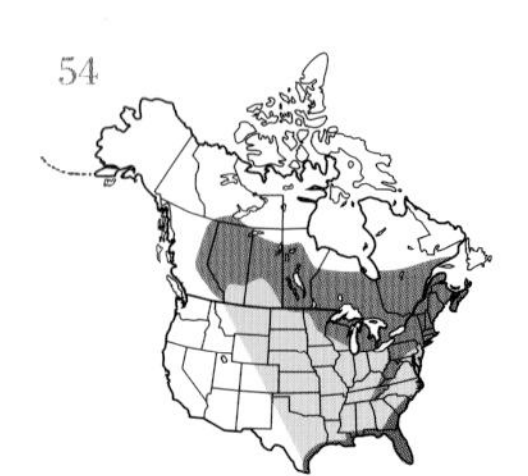

Blue-headed Vireo

Vireo solitarius

Although the sexes of Blue-headed Vireo are broadly similar, some males are noticeably more brightly marked than most females. All adults have a blue-gray hood, grading to olive-green on the nape and back. The throat is white and the underparts are otherwise pale with a buffish-yellow suffusion to the flanks. Note the striking white "spectacles" around the eyes. The dark wings have two very pale yellowish wingbars and margins to the inner flight feathers. The tail has white outer feathers. Juveniles are similar to adults but duller overall than a typical adult female.

The Blue-headed Vireo is present as a breeding species mainly from April to September, its range extending across much of northern temperate North America. At other times of the year its non-breeding range extends from southeast U.S.A. to Central America. Like many vireos, this species is unobtrusive, and silent birds are easily overlooked as they forage for insects in dappled foliage.

FACT FILE

LENGTH 5 in (12.5 cm)

FOOD Insects and other invertebrates

HABITAT Wide range of woodland habitats

STATUS Widespread and common summer visitor

VOICE Song comprises a series of well-spaced whistles. Call is a rasping *tche*. The species is a competent mimic of other vireo species

adult
life-size

Hutton's Vireo

Vireo huttoni

adult

Hutton's Vireo is a small, dumpy songbird with a relatively large head. At first glance it could be confused with a Ruby-crowned Kinglet (p.123), but the bill is stubby, not needle-like, and the calls and songs are very different. The sexes are similar. Adults have greenish-gray upperparts, brightest in birds from the Pacific coast and grayest in birds from the interior of the range. The underparts are pale greenish gray, palest on the throat. The dark eye is emphasized by the whitish partial eyering and pale lores. The dark wings have two white wingbars and pale margins to the inner flight feathers. The legs are blue-gray. Juveniles are similar to adults.

Hutton's Vireo is present year-round as a breeding species down the west coast of North America; it is also found in southern upland woodlands in Texas, New Mexico, and Arizona. The species is an altitudinal migrant in some parts of its range, moving to lower elevations outside the breeding season. It forages for insects in foliage and in winter it often associates with nomadic mixed flocks of other small songbirds.

adult life-size

FACT FILE

LENGTH 5 in (12.5 cm)

FOOD Insects and other invertebrates

HABITAT Evergreen woodlands, particularly where live oaks thrive

STATUS Locally common resident

VOICE Song comprises *tsu-wee* or *tsee-oo* phrases, repeated every second or so. Call is a rasping *rrhe*

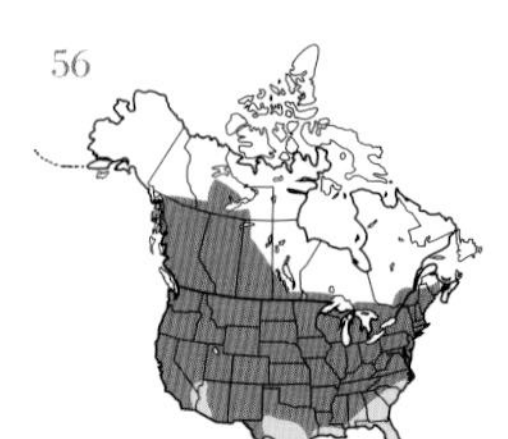

Warbling Vireo

Vireo gilvus

The Warbling Vireo recalls a plain warbler but has a relatively large head and stout bill. Given the species' geographical variation (see below), the sexes are similar. Adults have dull greenish-yellow upperparts, subtly darkest on the forecrown. The underparts are pale overall with a faint yellow suffusion to the flanks and undertail. The head has a whitish supercilium and lores, and the wings lack obvious wingbars. Eastern birds (subsp. *gilvus*) are overall paler and have longer bills than western birds (subsp. *swainsoni*). Juveniles are similar to their respective subspecies adults but with a more obvious yellow flush to the flanks and undertail.

The Warbling Vireo is present as a breeding species across much of temperate North America, mainly from April to September. It spends the rest of the year in Central America. It is an unobtrusive songbird and usually searches for insect prey high in the tree canopy. Its presence is often initially detected by hearing its distinctive song.

FACT FILE

LENGTH 5.5 in (14 cm)

FOOD Insects and other invertebrates

HABITAT Deciduous riverside woodland

STATUS Widespread and common summer visitor

VOICE Song comprises a series of warbling phrases, with distinct pauses in between. Call is a nasal *nrrr*

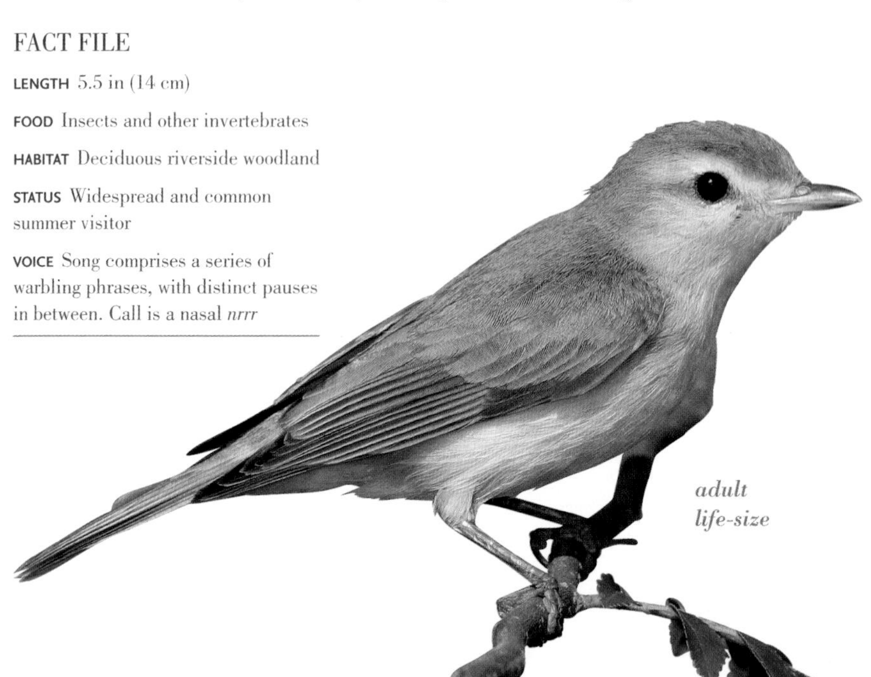

adult life-size

Philadelphia Vireo

Vireo philadelphicus

The Philadelphia Vireo recalls a plump-bodied warbler (notably the Tennessee Warbler; p.190), but has a stouter bill and bolder head pattern. The sexes are similar. Adults have an olive-green back and neck with a grayish crown. The underparts are pale overall but variably flushed yellow. The face pattern comprises a white supercilium and dark eye stripe with a white line below it. Juveniles are similar to adults but the yellow suffusion to the underparts is more striking.

The Philadelphia Vireo is present as a breeding species across northern North America, mainly from May to September. It spends the rest of the year in Central America. Its localized summer distribution reflects its preference for young-growth trees in a woodland mosaic of different age classes. Regrowth after woodland management or selective clearance appears to benefit the species. It is unobtrusive and easiest to detect by its song.

FACT FILE

LENGTH 5.25 in (13.5 cm)

FOOD Insects and other invertebrates

HABITAT New-growth deciduous woodland

STATUS Widespread but generally local summer visitor

VOICE Song recalls that of Red-eyed Vireo (p.58): a series of short phrases such as *tse-oo-wit*, *tsee-oo* with pauses between. Call is a nasal *tchrrr*

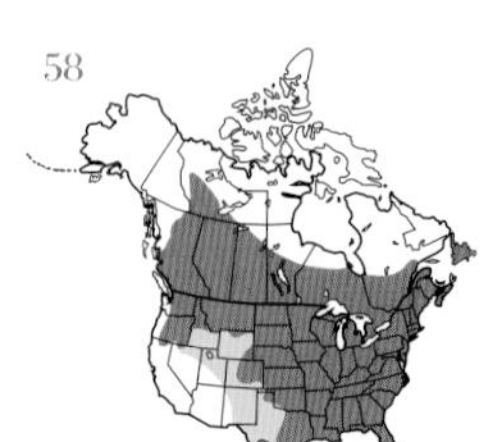

Red-eyed Vireo

Vireo olivaceus

This is North America's most familiar vireo. The sexes are similar. Adults have a greenish-gray back and neck, and rather grubby white underparts with a dull yellow flush to the flanks and undertail. The striking head pattern comprises a dark gray crown and long white supercilium, defined above and below by black lines. The eye has a red iris, and the bill is stout and relatively long. Juveniles are similar to adults but the iris color is subtly duller.

FACT FILE

LENGTH 6 in (15 cm)

FOOD Insects and other invertebrates

HABITAT Deciduous woodland

STATUS Widespread and common summer visitor

VOICE Song comprises a series of two- to four-syllable phrases, including *tse-oo-wit*, *tse-oo-ee*, and *tsee-oo*. Call is a nasal *zzNrrr*

adult life-size

adult

The Red-eyed Vireo is present as a breeding species across much of northern and eastern North America, mainly from May to August. It spends the rest of the year in South America. It usually forages for insects in an unobtrusive manner, making it hard to spot in dappled foliage; its presence is often detected first by its distinctive song.

adult life-size

Gray Jay

Perisoreus canadensis

The Gray Jay is a large-headed, plump-bodied songbird with soft-looking plumage. The sexes are similar. Adults have a dark gray back, separated from the dark gray rear crown by a pale gray nape. The rest of the face, and the underparts, are pale gray. The wings are dark, with subtly pale feather margins in most birds, and the tail is dark. Birds from the northwest of the species' range have the darkest upperparts and birds from the southwest are palest overall. Juveniles differ from adults in being uniformly dark gray.

The Gray Jay is a resident breeding species across northern North America and in western mountain ranges. Outside the breeding season it is often seen in small roving groups. Birds tend to be inquisitive and opportunistic feeders, so if you go camping within the species' range it is likely to come to you, on the lookout for food scraps.

adult life-size

FACT FILE

LENGTH 11.5 in (29 cm)

FOOD Wide range of foods, from seeds and berries to insects and campfire scraps

HABITAT Northern and upland conifer forests

STATUS Widespread and fairly common resident

VOICE Not especially vocal. Calls include a fluty, whistled *wheeo* and a harsh, chattering *chakk*

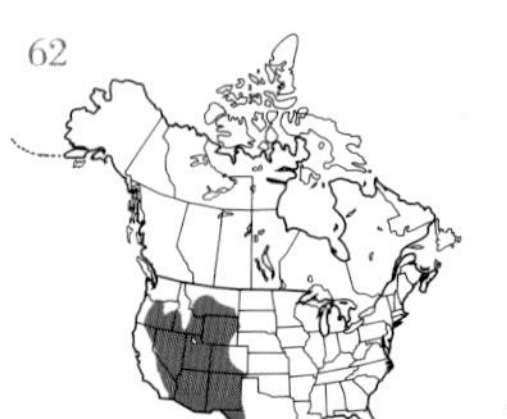

Pinyon Jay

Gymnorhinus cyanocephalus

The Pinyon Jay is a slim-bodied, slender-billed songbird. The sexes are similar. Adults are dull blue overall, darkest on the crown, back, and wings, and palest on the throat and belly. A close view reveals faint streaking on the throat. The bill is dark and the legs are blackish. Juveniles are uniformly pale blue-gray and much duller overall than adult birds.

adult
life-size

FACT FILE

LENGTH 10.5 in (26.5 cm)

FOOD Mainly pine seeds, which it stores, but also an opportunistic feeder (insects, fruit, seeds, etc.)

HABITAT Upland pinyon–juniper forests

STATUS Widespread and fairly common resident

VOICE Calls include a nasal, quail-like *huah-keeh-keeh*

The Pinyon Jay is a widespread but declining resident of mountain ranges in the Midwest region. If you see one Pinyon Jay then you are likely to see dozens: It nests colonially and outside the breeding season it gathers in flocks, these occasionally numbering 100 or more birds.

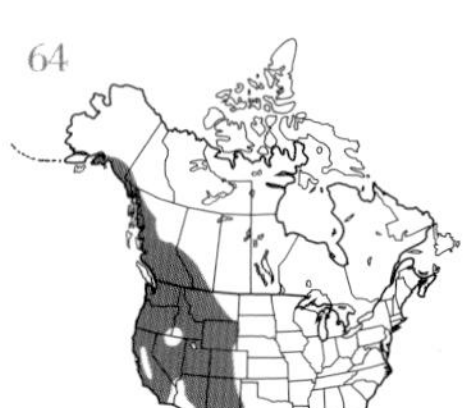

Steller's Jay

Cyanocitta stelleri

Steller's Jay is a colorful medium-sized songbird. The striking crest is longest in subspecies from the southern Rockies and shortest (also with some blue feathering) in Pacific coast subspecies. Given this regional variation, the sexes are similar. Adults have a dark sooty gray back and nape, and a mainly blackish head. The plumage is otherwise dark blue with variable amounts of dark laddering on the wings and tail. Juveniles are similar to adults but overall duller and browner.

FACT FILE

LENGTH 11.5 in (29 cm)

FOOD Omnivorous and opportunistic

HABITAT Conifer forests and mixed woodlands

STATUS Common resident

VOICE Calls include a chattering rattle and a harsh *scheck*

adult life-size

Steller's Jay is present year-round within its given western range, although in some years birds are irruptive, moving south and east outside the breeding season. The species is very inquisitive and will often visit campers and picnickers, on the lookout for scraps of food.

Blue Jay

Cyanocitta cristata

adult

The Blue Jay is a familiar colorful songbird with an obvious crest. The sexes are similar. Adults have a purplish-blue cap, nape, and back. The wings are mainly blue but with black barring, as well as bold white patches and a wingbar. The tail is blue with black barring and white outer tips. The underparts are pale gray-blue, with a dark "necklace" on the throat. Juveniles are similar to adults but less colorful.

The Blue Jay is present as a breeding species across most of eastern and central North America. Generally birds are sedentary in their habits but in some years northern populations in particular undertake mass irruptive movements south. The species is a frequent visitor to garden birdtables and is usually bold and inquisitive in locations such as parks.

FACT FILE

LENGTH 11 in (28 cm)

FOOD Omnivorous and opportunistic

HABITAT Wide range of wooded habitats, including gardens

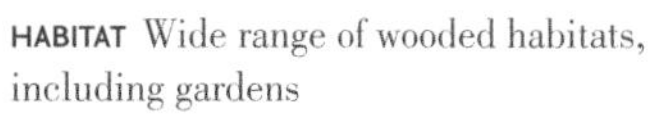

STATUS Widespread and common resident

VOICE Calls include a shrill *jay- jay- jay*, a whistling *pee-de-de*, and mimicry, especially of raptors

adult life-size

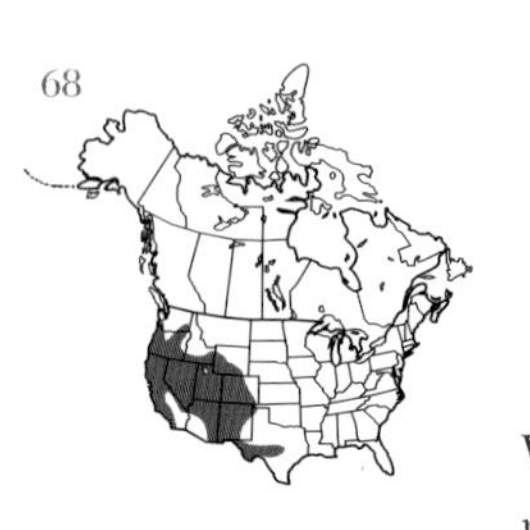

Western Scrub-Jay

Aphelocoma californica

Within its North American range, the Western Scrub-jay is represented by an array of subspecies, which can be lumped into two main groups: Pacific coast birds and interior birds. Within these groups, the sexes are similar. Adult Pacific birds have mainly blue upperparts except for the gray-brown back, dark cheeks, and faint white supercilium. The underparts are pale gray except for the streaked whitish throat. Adult interior birds are similar but the upperparts are paler and brighter. Juveniles recall their adult counterparts but are overall dull gray above, with a blue tail and flight feathers, and pale gray underparts.

adult

FACT FILE

LENGTH 11 in (28 cm)

FOOD Omnivorous and opportunistic

HABITAT Wide range of wooded habitats

STATUS Common resident

VOICE Utters a variety of harsh and chattering calls, including a rasping *cheerp*

adult life-size

The Western Scrub-jay is present year-round, and mainly sedentary, in its southwestern range in North America. Pacific coast birds are usually bold and inquisitive, making them easy to see. Interior birds tend to be much shyer and skulking.

adult

Clark's Nutcracker

Nucifraga columbiana

Clark's Nutcracker is a well-marked and distinctive songbird. The sexes are similar. Adults have subtle pale purplish-gray body plumage that is palest on the face and undertail. The wings are mainly black except for a broad white patch (the inner flight feathers). The tail has black central feathers and white outer feathers. The legs and slender bill are black. Juveniles are similar to adults but subtly paler overall.

adult life-size

FACT FILE

LENGTH 12 in (30.5 cm)

FOOD Mainly conifer seeds, but also an opportunistic feeder

HABITAT Mountain conifer forests

STATUS Widespread and common resident

VOICE Calls include a rasping *shrerr-shrerrr* and harsh *craa-craa-craa*

Clark's Nutcracker is present year-round in its favored upland terrain. Pine nuts are the mainstay of the species' diet, and the chances of individual birds surviving the harsh winter months are boosted by the creation of large stores of nuts, used as a larder. There is also an inquisitive side to their nature, and birds on the lookout for food will sometimes visit forest campers and picnickers.

Black-billed Magpie

Pica hudsonia

The Black-billed Magpie is a distinctive, mainly black and white songbird with a long tail and short, rounded wings. The sexes are similar. Adults have mainly black plumage with a contrasting white belly and white patch on the closed wing; the latter appears as a striking white wing patch in flight. In the right light, there is a bluish-green sheen to the wings and tail. Juveniles are similar to adults but have a pale (not dark) eye, shorter tail, and duller plumage that lacks any iridescence.

adult

FACT FILE

LENGTH 19–21 in (48–53.5 cm)

FOOD Omnivorous and opportunistic feeder

HABITAT Wide range of lightly wooded and open habitats

STATUS Widespread and common resident

VOICE Calls include a range of harsh, screeching calls and a chattering *chek-chek-chek*

The Black-billed Magpie is present year-round within its range in western North America. On the ground, it walks with a characteristic swagger. Outside the breeding season it often gathers in flocks. The species' varied diet means it is quick to exploit an opportunity, and this can include scavenging at roadkill carcasses and raiding discarded trash in urban settings.

adult
life-size

American Crow

Corvus brachyrhynchos

The American Crow is the most widespread bird of its kind in the region and the default corvid to which its relatives should be compared. The sexes are similar. Adults have uniformly glossy black plumage. The bill is proportionately long and dark, and the legs are dark. In flight, the relatively long tail is sometimes fanned. Juveniles are similar to adults but the eye has a pale (not dark) iris and there is a subtle brown tinge to the plumage. The closely related **Northwestern Crow** (*C. caurinus*) is essentially identical to American Crow populations found in the northwest; in areas where the two species do not occur together, geographical range represents the only realistic prospect of separating them.

adult Northwestern Crow

FACT FILE

LENGTH 15–18 in (38–45.5 cm)

FOOD Omnivorous and opportunistic

HABITAT Wide range of habitats, from farmland to urban environments

STATUS Widespread and common resident and partial migrant

VOICE Call is a raucous *caaw-caaw*

adult

adult
life-size

The American Crow is resident year-round across most of North America, except along the northwest coast, and in the far north, where populations migrate south in fall. All manner of food items are eaten and the species is quick to cash in on human wastefulness, visiting garbage dumps and raiding trash cans. It sometimes gathers in flocks outside the breeding season. The Northwestern Crow replaces the American Crow along the northwest coast of British Columbia and Alaska.

Fish Crow

adult

Corvus ossifragus

The Fish Crow is superficially similar to the American Crow (p.74) and their ranges overlap. Subtle plumage and structural differences do exist, but their calls provide the best chance of separating the two. The sexes are similar. Adults have glossy, uniformly black plumage, subtly more shiny than in American Crow. The bill is long, subtly longer in relative terms than that of American Crow. In flight, compared to American Crow, the wings are more pointed and the tail is longer and more fan-shaped. Juveniles are similar to adults but with a brown tinge to the plumage.

FACT FILE

LENGTH 19 in (48 cm)

FOOD Omnivorous and opportunistic feeder

HABITAT Shores, estuaries, and freshwater wetlands

STATUS Widespread and common resident

VOICE Call is a *cah-haah*, more nasal than that of American Crow (p.74)

adult

The Fish Crow is a year-round resident in southeastern U.S.A. The vast majority of birds are found near the coast, favoring shorelines and estuaries for foraging. Fish Crows do also occur inland, although they are invariably seen in the vicinity of water, feeding on the shores of lakes and rivers. The species is sociable, often seen in flocks, especially outside the breeding season.

adult
life-size

Common Raven

Corvus corax

The Common Raven is the largest songbird in North America. The sexes are similar, and adults have a uniformly black plumage with an iridescent, oily-looking sheen. The bill is massive and dark, the legs are black, and the throat appears shaggy. Juveniles are similar to adults but the eye has a paler iris and there is a brownish tinge to the plumage. In flight, all birds have a long, wedge-shaped tail. The **Chihuahuan Raven** *C. cryptoleucus* is slightly smaller (Length 20 in/51 cm) but virtually identical in the field.

The Common Raven is a year-round resident, typically seen in pairs although larger groups gather in winter when the feeding is good. The diet is varied but the species often scavenges at carcasses in winter. Common Ravens are very aerobatic in flight, able to soar with ease but often tumbling, twisting and turning in mid-air. The Chihuahuan Raven is a resident of arid southwestern habitats. Outside the breeding season it often forms large, nomadic flocks.

adult

FACT FILE

LENGTH 24–25 in (61–63.5 cm)

FOOD Omnivorous and opportunistic feeder

HABITAT Wide range of habitats, from tundra and upland forests to farmland

STATUS Widespread and common resident

VOICE Call is a deep, resonant *cronk*

Horned Lark

Eremophila alpestris

The Horned Lark gets its name from the small black projecting feathers on its head. The sexes are similar overall, although males are more strikingly marked on the face than females. Adults have brown upperparts and whitish underparts with a dark breast band. The head has a black mask, forecrown, and "horns" on the side of the crown; the face is otherwise yellowish. Subspecies variation affects the intensity of color on the face and on the back. Juveniles have speckled upperparts and show a suggestion of the adult's facial markings. In flight, all birds show pale underwings.

adult life-size

The Horned Lark is a widespread species and present year-round across much of the U.S.A. In winter, populations in the south of the species' range are swollen by birds that have moved south from northern and Arctic summer breeding grounds to escape the harsh winters. The species spends much of the time feeding unobtrusively on the ground, creeping along with its body held low. It forages for insects in summer but feeds mainly on seeds in winter. It often forms flocks outside the breeding season.

FACT FILE

LENGTH 7–8 in (18–20 cm)

FOOD Insects and seeds

HABITAT Open barren habitats, ranging from tundra to sparse grassland

STATUS Common, northern populations are migratory; more sedentary in south of range

VOICE Song comprises a few rasping *chrrt* notes, followed by a series of tinkling notes. Flight calls include a thin *tsee-titi*

adult

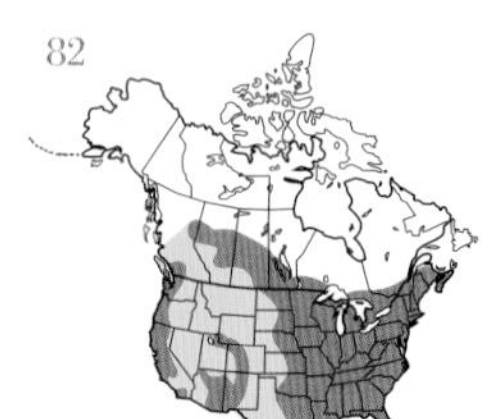

Purple Martin

Progne subis

The Purple Martin is a familiar songbird in North America. The sexes differ. Adult males are uniformly dark, with a purplish sheen visible in good light. Adult females have gray-brown upperparts with a suggestion of a bluish sheen, and mottled gray-brown underparts, palest as a patch on the belly. Juveniles are similar to an adult female but with much paler underparts. Males in their first spring of life retain a variable extent of juvenile plumage features. In flight, all birds have broadly triangular wings and a relatively long, forked tail.

The Purple Martin is present as a breeding species mainly from April to August; it is widespread in eastern North America but more scattered elsewhere. It spends the rest of the year in South America. In summer, the species is often associated with human settlement and responds well to the introduction of nestboxes. It catches flying insects on the wing.

male
life-size

FACT FILE

LENGTH 7–8 in (18–20 cm)

FOOD Insects

HABITAT Suburban habitats and open country

STATUS Locally common summer visitor

VOICE Song comprises a series of gurgling and croaking notes. Calls include a liquid *chrrr* and various whistles

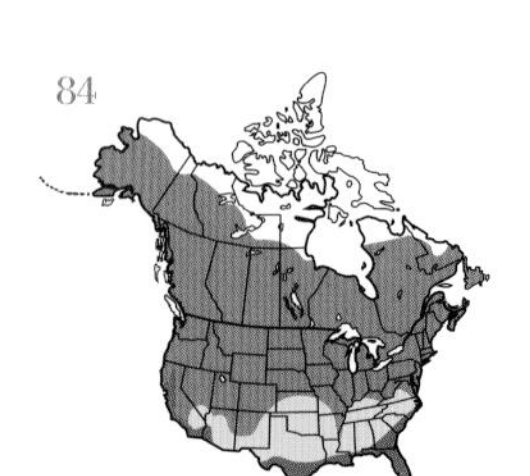

Tree Swallow

Tachycineta bicolor

The Tree Swallow has contrastingly bicolored plumage. The sexes are subtly dissimilar. Adult males and many females have blackish upperparts, with a blue sheen seen in good light, and demarcated as a distinct dark cap on the head. The underparts are white. Some females, and most adults in fall, have browner upperparts with little or no sheen. Juveniles are similar to a dull adult, sometimes with a hint of a gray breast band. In flight, all birds have triangular wings and a slightly forked tail.

The Tree Swallow is present as a breeding species mainly from April to September. Most birds head for Central America for the rest of the year, although small numbers winter across southern U.S.A. The species catches insects in flight and sometimes gathers in flocks if the feeding is good; when times are hard, it will also eat berries in winter. It nests in treeholes but will also use suitable nestboxes.

female

male
life-size

FACT FILE

LENGTH 5.5–6 in (14–15 cm)

FOOD Mainly insects, but also berries

HABITAT Open habitats, especially wetlands, with dead trees

STATUS Widespread and common summer visitor

VOICE Call and song comprise a series of whistling chirps

Violet-green Swallow

Tachycineta thalassina

The Violet-green Swallow recalls a Tree Swallow (p.84) but has a white face (the dark cap extends below the eye in Tree Swallow) and striking white sides to the rump. The sexes are similar. Adults have a greenish back and wing coverts, a greenish-brown cap, and a violet sheen to the wings and uppertail. The underparts are white. A few females have browner upperparts, as do all juveniles, which are duller still, with a grubby gray face. In flight, all birds have narrower, longer wings than Tree Swallow; these extend beyond the tail in perched birds.

The Violet-green Swallow is present as a breeding species mainly from April to September. It spends the rest of the year in Central America. It catches insects on the wing and gathers in flocks where the feeding is good. The species nests in treeholes.

female

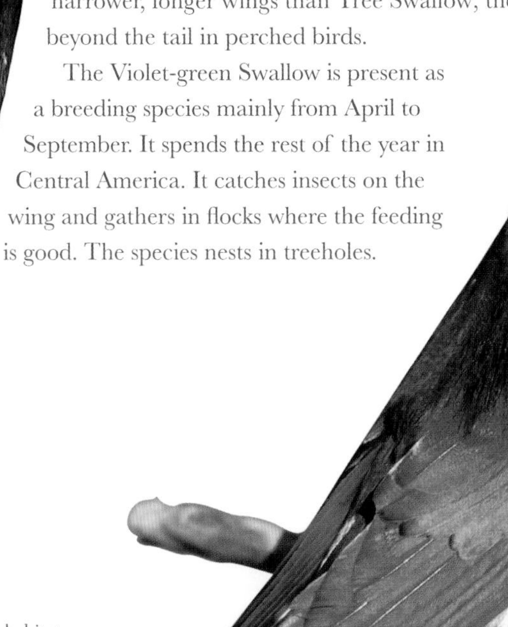

male
life-size

FACT FILE

LENGTH 5.5 in (14 cm)

FOOD Insects

HABITAT A range of open habitats, from coasts and deserts to lightly forested mountain slopes

STATUS Widespread and common summer visitor

VOICE Song and calls comprise a range of twittering and whistling notes

Northern Rough-winged Swallow

Stelgidopteryx serripennis

The Northern Rough-winged Swallow is similar to the Bank Swallow (p.87) but lacks that species' dark breast band, and it has grubbier underparts. The sexes are similar. Adults have brown upperparts with pale fringes to the inner flight feathers and wing coverts. The throat and breast are pale gray-buff, grading to whitish on the rest of the underparts. Juveniles are similar to adults but with more rusty-brown upperparts, buff on the throat and breast, and rufous-buff margins to the inner flight feathers and wing coverts. In flight, all birds have narrow, pointed wings and an almost square-cut end to the tail.

adult

The Northern Rough-winged Swallow is present as a breeding species across much of temperate North America, mainly from April to September. It spends the rest of the year in Central America. The species nests in rock crevices and holes in banks, but never in large colonies like the Bank Swallow. It catches insects on the wing and often feeds over water.

adult life-size

FACT FILE

LENGTH 5 in (12.5 cm)

FOOD Insects

HABITAT Open habitats, usually in the vicinity of sand banks or cliffs for nesting

STATUS Widespread and common summer visitor

VOICE Song and calls comprise a range of buzzing notes

Bank Swallow

Riparia riparia

The Bank Swallow is a small songbird that spends much of its time on the wing. The sexes are similar. Adults have gray-brown upperparts and mainly white underparts with a clearly defined brown breast band. There is a clear-cut division between the brown cap and white throat. Juveniles are similar to adults but the wing feathers have pale rufous margins. In flight, all birds show rather narrow wings and a slightly forked tail.

adult

The Bank Swallow is present as a breeding species across much of northern North America, mainly from April to September. It spends the rest of the year in South America. It catches insects in flight, often over water, gathering in considerable numbers if the feeding is good. As their name suggests, Bank Swallows nest in holes in banks, often in sizeable colonies.

adult
life-size

FACT FILE

LENGTH 5.25–5.5 in (13.5–14 cm)

FOOD Insects

HABITAT Open wetland habitats, with banks for nesting

STATUS Widespread and common summer visitor

VOICE Song comprises a series of abrupt, twittering notes. Typical call is a buzzing *prrrt*

Cliff Swallow

Petrochelidon pyrrhonota

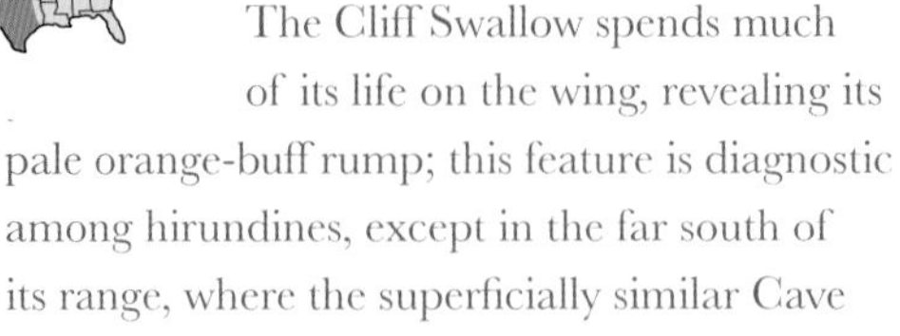

adult

The Cliff Swallow spends much of its life on the wing, revealing its pale orange-buff rump; this feature is diagnostic among hirundines, except in the far south of its range, where the superficially similar Cave Swallow (p.89) also occurs. The sexes are similar. Adults have a dark cap that is separated from the pale-streaked bluish-black back by a reddish collar. This color extends to the face, becoming dark on the throat. The forehead is white and the buffish rump contrasts with the square-ended dark tail. The underparts are mainly pale but with darker spots on the undertail. Juveniles are much less colorful and less strikingly marked than adults, with a rather dark head and throat. In flight, all birds show broadly triangular wings and a square-ended tail.

The Cliff Swallow is present as a breeding species across much of North America, mainly from April to September. It spends the rest of the year in South America. It catches insects on the wing and nests colonially, building nests on cliffs and manmade structures such as bridges.

FACT FILE

LENGTH 5.5 in (14 cm)

FOOD Insects

HABITAT Range of habitats, from open country to forests; anywhere with cliffs for nesting

STATUS Widespread and common summer visitor

VOICE Song and calls comprise a range of soft twittering notes

adult
life-size

Cave Swallow

Petrochelidon fulva

The Cave Swallow is similar to the Cliff Swallow (p.88); distinguishing features include the reddish (not white) forehead and paler throat and face. The sexes are similar. Adults have a dark bluish cap, separated from the white-streaked dark bluish back by a pale reddish collar. This color extends to the cheeks and throat. The forehead is reddish, and the reddish-buff rump contrasts with the square-ended dark tail. The underparts are mostly pale but with darker spots on the undertail. Juveniles are less colorful and less strikingly marked than adults, with a pale buff forehead, grubby-looking nape, and whitish throat.

The Cave Swallow is present as a breeding species, mostly to southern Texas, and mainly from March to August. Small numbers remain in winter, but most head for Central America for the rest of the year. As its name suggest, the Cave Swallow builds mud nests in caves, but also under bridges. It catches insects on the wing and often feeds over water.

adult
life-size

FACT FILE

LENGTH 5.5 in (14 cm)

FOOD Insects

HABITAT Open country, cliffs, and gorges

STATUS Locally common summer visitor

VOICE Song includes various twittering notes, and call-like *che-wiit* phrases

Barn Swallow

Hirundo rustica

The Barn Swallow is a familiar songbird, often seen in flight or perched on overhead wires. The sexes are subtly dissimilar. Adult males have a blue cap, nape, and back, with a red forehead and throat. A dark breast band separates the throat color from the buffish-orange underparts. The tail has long streamers (extensions to the outer feathers) that are very obvious in flight. Adult females are similar but have much paler underparts and shorter tail streamers. Juveniles are similar to an adult female but with even shorter tail streamers and a dull buff throat and forehead.

adult

The Barn Swallow is present as a breeding species across much of North America, mainly from March to September. It spends the rest of the year in South America. It builds a mud nest, sometimes in a natural setting such as a cave, but often on a ledge in an agricultural barn or shed. It chases insects in flight and often feeds over water.

adult life-size

FACT FILE

LENGTH 6.5–7 in (16.5–18 cm)

FOOD Insects

HABITAT Open country, including grassland and farmland

STATUS Widespread and common summer visitor

VOICE Song comprises a series of twittering warbles, and grating notes. Calls include a sharp *che-viit*

Carolina Chickadee

Poecile carolinensis

The Carolina Chickadee is very similar to its Black-capped Chickadee cousin (p.92) and, in the limited zone where the two species overlap, certain identification may not always be possible. The sexes are similar and both adults and juveniles have a gray-buff back. The dark wings have pale margins to the inner flight feathers and greater coverts, but these features are less striking than in Black-capped. The head has a black cap, throat, and bib (like Black-capped), and white cheeks that grade to pale gray on the sides of the nape. The underparts are overall pale gray with a faint buff suffusion on the flanks; the result is that the flanks are subtly "warmer" in tone than in Black-capped. The dark tail has pale feather margins. The legs are blue-gray and the bill is dark.

The Carolina Chickadee is present year-round in wooded habitats in southeast U.S.A. In the wild, it nests in treeholes but, like many related songbirds, it will happily use a nestbox. It also visits garden feeders, especially in winter, and outside the breeding season will join nomadic mixed-species songbird flocks as they search for food.

FACT FILE

LENGTH 4.75 in (12 cm)

FOOD Invertebrates and seeds

HABITAT Deciduous woodland

STATUS Widespread and common resident

VOICE Song is a four-note whistling *fee-bee fee-bay*. Call is rapid *chika-dee-dee*, subtly higher in pitch than that of Black-capped Chickadee (p.92)

adult life-size

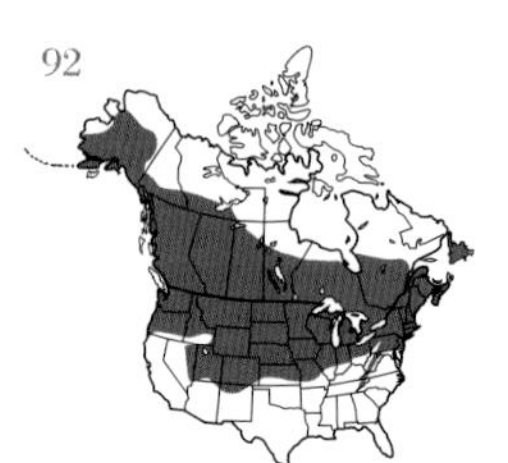

Black-capped Chickadee

Poecile atricapillus

The Black-capped Chickadee is a familiar songbird. The sexes are similar, and both adults and juveniles have a gray-buff back. The wings are mostly dark but have a pale panel created by whitish margins to the inner flight feathers and greater coverts. The cheeks are white, and this color extends to the sides of the nape. The head has a black cap that tapers down the nape. There is a black throat and bib, and the underparts are otherwise pale with a subtle pinkish-buff suffusion to the rear of the flanks. The dark tail has pale feather margins, the legs are blue-gray, and the bill is dark.

adult
life-size

The Black-capped Chickadee is present year-round across much of central North America. It favors woodland but is a familiar visitor to garden feeders, especially outside the breeding season. Although generally sedentary, it often consorts with nomadic mixed-species songbird flocks in winter. In the wild, Black-capped Chickadees nest in treeholes, but they will happily use artificial nestboxes.

FACT FILE

LENGTH 5.25 in (13.5 cm)

FOOD Invertebrates and seeds

HABITAT Wide range of wooded habitats

STATUS Widespread and common resident

VOICE Song is a whistled, disyllabic *fee-bee*. Onomatopoeic call is *chika-dee-dee-dee*

adult

Mountain Chickadee

Poecile gambeli

The Mountain Chickadee recalls its Black-capped Chickadee cousin (p.92) but the diagnostic white supercilium allows certain separation. The sexes are similar, as are adults and juveniles. The back is gray in birds in the south of the species' range but gray-buff in birds in the north. All birds have pale fringes to the inner flight feathers and coverts, creating a subtle panel. The black cap has a white supercilium, and the throat and bib are black. The pale underparts have a gray wash to the flanks. The legs are blue-gray and the bill is dark.

As its name suggests, the Mountain Chickadee is a high-altitude species, present year-round in its western North American range. It occurs up to the treeline in summer but may descend to lower levels in winter, when it often joins roving mixed-species flocks of small songbirds. Conifer seeds are stored as a larder for the winter months.

adult life-size

FACT FILE

LENGTH 5.25 in (13.5 cm)

FOOD Invertebrates and seeds

HABITAT Mountain conifer forests

STATUS Locally common resident

VOICE Song is a four-note *fee-bee, fee-bay*. Call is a nasal *chika-tzee-tzee*

Chestnut-backed Chickadee

Poecile rufescens

adult

The Chestnut-backed Chickadee's colorful plumage (by chickadee standards) makes identification relatively easy. The sexes are similar, as are adults and juveniles. All birds have a distinctive chestnut back. The dark cap and black throat frame the white cheeks, and the darkish wings have pale feather margins, giving them a silvery look. The underparts are overall gray-buff in Californian birds but heavily flushed with chestnut on the flanks in birds found elsewhere in the species' range. The legs are blue-gray and the bill is dark.

Within its restricted, mainly coastal, range the Chestnut-backed Chickadee is present year-round, favoring areas of tall mature conifers, including rainforests in the Pacific Northwest. Outside the breeding season it is often found in roving mixed-species flocks of small songbirds.

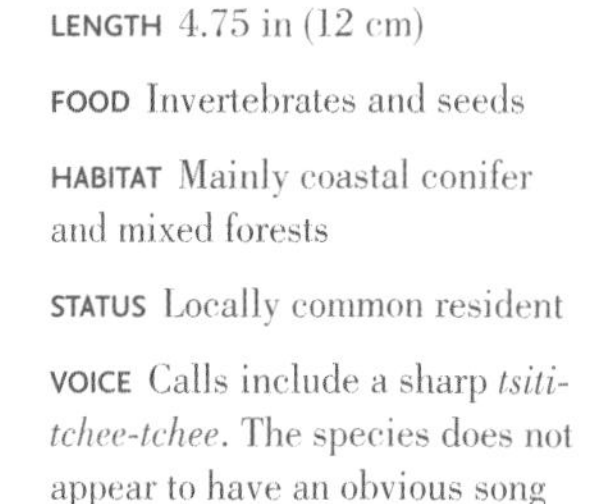

FACT FILE

LENGTH 4.75 in (12 cm)

FOOD Invertebrates and seeds

HABITAT Mainly coastal conifer and mixed forests

STATUS Locally common resident

VOICE Calls include a sharp *tsiti-tchee-tchee*. The species does not appear to have an obvious song

adult
life-size

Boreal Chickadee

Poecile hudsonicus

The Boreal Chickadee is a distinctive little songbird. The sexes are similar, as are adults and juveniles. All birds have a brown cap that tapers down the nape and joins the brown back. The wings are dark with faint pale feather margins. The face is white at the front, grading to gray on the cheeks, and the throat is black. The pale underparts are suffused with orange-buff on the flanks.

The Boreal Chickadee is present year-round in its northern and Arctic range. Although widespread, it is seldom numerous and typically territories are thinly scattered. To survive the harsh northern winters it relies on larders of stored seeds, collected in summer and fall. Although usually sedentary, populations are sometimes forced to move south if the seed crops fail.

adult

adult life-size

FACT FILE

LENGTH 5.5 in (14 cm)

FOOD Invertebrates and seeds

HABITAT Northern boreal forests

STATUS Widespread and fairly common resident

VOICE Song is a vibrant trill. Calls include a thin *dsee* and a *tsika-day-day*

Oak Titmouse

Baeolophus inornatus

The Oak Titmouse has rather plain, nondescript plumage. This absence of obvious plumage features allows separation from all other titmouse species except Juniper Titmouse (p.97). The sexes are similar, as are adults and juveniles. All birds have unmarked gray-brown upperparts; the pale gray underparts are flushed pinkish buff on the flanks. The head has a short crest and stubby bill, and the legs are dark.

The Oak Titmouse is present year-round in woodland on the Pacific slopes of California and Oregon. Outside the breeding season it is often found in small groups, sometimes associating with nomadic mixed-species flocks of small songbirds. The species is very similar to Juniper Titmouse, but overall it is subtly darker and browner; the two species' ranges barely overlap, and their habitat preferences differ.

adult life-size

FACT FILE

LENGTH 5 in (12.5 cm)

FOOD Invertebrates and seeds

HABITAT Wooded habitats, especially where oaks predominate

STATUS Widespread and locally common resident

VOICE Song comprises various whistling phrases, including *peechew*. Calls include a sharp *tsita-chrr*

adult

Juniper Titmouse

Baeolophus ridgwayi

The Juniper Titmouse is similar to, and subtly larger than, Oak Titmouse (p.96). The sexes are similar, as are adults and juveniles. All birds have pale gray upperparts, and even paler gray underparts that are sometimes very subtly suffused buff on the flanks. The pointed bill is marginally longer than in Oak Titmouse, although this feature is of little use for identification unless you are very familiar with both species. The legs are blue-gray.

adult

The Juniper Titmouse is present year-round, favoring specialized upland habitats that are very different from those frequented by Oak Titmouse; the ranges of the two species barely overlap. Outside the breeding season, Juniper Titmouse is often found in small, roaming groups, and sometimes consorts with larger mixed-species flocks.

FACT FILE

LENGTH 5.25–5.5 in (13.5–14 cm)

FOOD Invertebrates and seeds

HABITAT Open juniper (*Juniperus* spp.) woodland, often on mountain slopes

STATUS Widespread and fairly common resident

VOICE Song is varied, often including a three-phrase *whidla-whidla-whidla*; subtly lower in pitch than that of Oak Titmouse (p.96). Call is a sharp *tsika-chrr*

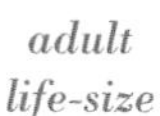

adult life-size

Tufted Titmouse

Baeolophus bicolor

With its hallmark crest, the Tufted Titmouse is a familiar and distinctive songbird. The sexes are similar. Adults have pale blue-gray upperparts with a blackish forehead and peaked crest. The very pale gray underparts are palest on the undertail, and suffused with orange-buff on the flanks. The pale face shows off the dark eye. The bill is dark and legs are blue-gray. Juveniles are similar to adults but the forehead is gray.

The Tufted Titmouse is a year-round resident of wooded habitats in eastern North America. It is not shy, making it easy to find and observe well. The species often visits garden feeders, especially in winter, and responds well to the introduction of nestboxes; these are a substitute for the treehole nest sites it uses in the wild.

adult

FACT FILE

LENGTH 6–6.5 in (15–16.5 cm)

FOOD Invertebrates and seeds

HABITAT Open deciduous woodland, and wooded parks and gardens

STATUS Widespread and common resident

VOICE Song comprises a series of disyllabic *pee-tu* phrases. Call is a nasal *zee-zee*

adult
life-size

Black-crested Titmouse

Baeolophus atricristatus

Once considered a subspecies of Tufted Titmouse (p.98), the Black-crested Titmouse is a distinctive songbird. The sexes are similar. Adults have mainly pale blue-gray upperparts with a pale forehead and black crown and crest. The very pale gray underparts are palest on the undertail and suffused with an orange-buff wash on the flanks. The pale face emphasizes the dark eye. The bill is dark and the legs are blue-gray. Juveniles lack the adults' black crown and crest, making them very similar to a juvenile Tufted Titmouse.

The Black-crested Titmouse is a year-round resident of woodland in southern U.S.A., including riverside forests and wooded parks and gardens. It nests in treeholes in the wild but responds well to the provision of nestboxes.

FACT FILE

LENGTH 6–6.5 in (15–16.5 cm)

FOOD Insects and seeds

HABITAT Open deciduous woodland

STATUS Widespread and common resident

VOICE Song comprises a series of *chiu* phrases. Call is a series of harsh *zree* notes

adult
life-size

Verdin

Auriparus flaviceps

With its slender, pointed bill and compact body, the Verdin bears a passing resemblance to a warbler. In overall plumage terms the sexes are similar, although the male's plumage is typically brighter than that of the female. Adults have a gray back and nape; the gray wings have pale margins to the flight feathers and a small reddish "shoulder" patch. The mainly yellow face emphasizes the dark lores and eye. The underparts are pale gray, and the legs and bill are dark. Juveniles are uniformly gray-buff above and pale gray below, and lack the colors seen on the adults' face and "shoulders."

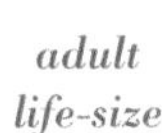

adult
life-size

The Verdin is present year-round in the desert habitats of the southern states of the U.S.A. Although it is an active feeder, flicking its tail as it progresses, it is usually unobtrusive and easily overlooked. The fact that it is usually solitary outside the breeding season does not help with locating the species.

juvenile

FACT FILE

LENGTH 4.5 in (11.5 cm)

FOOD Invertebrates

HABITAT Arid habitats, including deserts and mesquite woodland

STATUS Locally common resident

VOICE Song is a whistled three-note *tee-tiu-tiu*. Calls include a sharp *tsip*

Bushtit

Psaltriparus minimus

In plumage terms, there is subtle subspecies variation across the Bushtit's North American range. Given this geographical variation the sexes are similar, although adult females have a pale eye, while in males it is dark. Adults from the interior have blue-gray upperparts that are darkest on the primaries and tail, and pale gray underparts. The ear coverts are buff, and the legs and bill are dark. Adults from the Pacific coast are similar but have a brown cap. Juveniles are similar to their respective adults, although females have a dark eye for a few weeks. Some juvenile males in southwest U.S.A. have a black "mask;" this feature does not persist into adulthood.

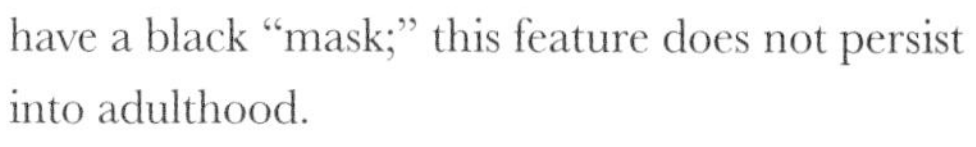

The Bushtit is present year-round in suitable habitats in western North America. Outside the breeding season it is usually found in small roving flocks, which are both active and vocal.

adult life-size

interior race

adult Pacific coast race

FACT FILE

LENGTH 4.5 in (11.5 cm)

FOOD Invertebrates

HABITAT Woodland and scrub

STATUS Widespread and common resident

VOICE A variety of buzzing and *tssip* notes that typically serve as contact calls

Red-breasted Nuthatch

Sitta canadensis

The Red-breasted Nuthatch is a plump-bodied forest songbird. The sexes are subtly different. Adult males have a blue-gray back, wings, and tail, and mainly reddish-buff underparts. The head has a black crown and eye stripe, and a white supercilium, face, and chin. Adult females are similar, but the black elements of the head plumage are paler and the underparts are less colorful. Juveniles are similar to their respective adults, but have brown, not black, wing feathers.

female

The Red-breasted Nuthatch is present year-round across the central part of its North American range, as well as in the west. However, northern populations move south in fall and in winter the species' range extends to most of the southern states. Outside the breeding season, Red-breasted Nuthatches often join roving flocks of mixed-species songbirds.

male
life-size

FACT FILE

LENGTH 4.5 in (11.5 cm)

FOOD Invertebrates and seeds

HABITAT Conifer forests

STATUS Widespread and common resident, and partial migrant

VOICE Song comprises a series of nasal *errn* notes, similar to the call, which is reminiscent of the sound of a child's toy trumpet

White-breasted Nuthatch

Sitta carolinensis

The White-breasted Nuthatch is the largest bird of its kind in the region. The sexes can be separated if seen together. Adult males have a blue-gray back and wings, with contrasting dark centers and pale edges to the tertials and wing coverts. The white face and throat contrast with the black nape and crown. The underparts are otherwise very pale gray, flushed rufous and white on the undertail. Adult females are similar but the crown and nape are dark gray. Geographical variation occurs across the species' range, represented by several subspecies. Eastern populations have paler gray backs and show more contrast in the wing markings than those from the west. Juveniles are similar to their respective adults but the wing feathers have buff fringes.

adult life-size

The White-breasted Nuthatch is present year-round across its extensive North American range, although it occasionally wanders in winter in response to food shortages. It is a regular visitor to garden feeders and sometimes joins roving mixed-species flocks of songbirds outside the breeding season.

adult

FACT FILE

LENGTH 5.75 in (14.5 cm)

FOOD Invertebrates and seeds

HABITAT Deciduous and mixed woodland

STATUS Widespread and common resident

VOICE Song is a series of nasal, whistling notes. Call is a nasal *nYen*

Pygmy Nuthatch

Sitta pygmaea

flight

adult

The Pygmy Nuthatch is a compact, dumpy-bodied forest songbird with a relatively large head and short tail. The sexes are similar. Adults have a blue-gray back and wing coverts, and subtly darker feathers. The cap is dull blue-green, defined below by a dark eye stripe; there is a pale patch on the nape. The face and throat are whitish, and the underparts are otherwise buff with a subtle blue-gray suffusion to the flanks. Juveniles are similar to adults but less colorful.

The Pygmy Nuthatch is present year-round in a range that is defined by the presence of Ponderosa Pines (*Pinus ponderosa*) and related species. It is an extremely social species. Cooperative behavior occurs when nesting, with breeding pairs often assisted by male helpers. Outside the breeding season, it forages in sizeable and vocal flocks, and even roosts communally.

FACT FILE

LENGTH 4.25 in (11 cm)

FOOD Invertebrates and seeds

HABITAT Conifer forests

STATUS Widespread and common resident

VOICE Song comprises a series of two-note *ke-Dee, ke-Dee…* phrases. Calls include endlessly repeated *kip-kip-kip…* notes

adult life-size

Brown-headed Nuthatch

Sitta pusilla

adult

The Brown-headed Nuthatch is a dumpy-bodied forest bird, and the southeastern counterpart of the Pygmy Nuthatch (p.104). The sexes are similar. Adults have a blue-gray back, tail, and wings, with subtly paler fringes to the edges of the inner flight feathers. The cap is brown, the brown nape has a bold whitish spot at the rear, and the face and throat are white. The underparts are overall pale but with a blue-gray wash to the flanks and a faint buff suffusion on the breast and undertail. Juveniles are similar to adults but paler overall and less colorful.

The Brown-headed Nuthatch is present year-round in a southeastern range that is defined by the presence of mixed-age native pines, with standing dead wood for nesting. It is an active feeder, climbing headfirst down tree trunks (like other nuthatches) and searching for insects on the outermost twigs.

FACT FILE

LENGTH 4.5 in (11.5 cm)

FOOD Invertebrates and seeds

HABITAT Pine forests

STATUS Widespread and common resident

VOICE Calls include various chirps and a nasal *Ke-waa*, like the sound of a child's squeaky toy; these vocalizations may also serve as a song

adult life-size

Brown Creeper

Certhia americana

With its downcurved needle-like bill and streaked brown plumage, the Brown Creeper is hard to mistake for any other species. The sexes are similar. Adults have brown upperparts adorned with pale teardrop spots, and whitish underparts suffused with buff on the flanks and undertail. The face has pale-streaked brown cheeks and a whitish supercilium. The short wings have buff barring and the rump and base of the tail are rufous. Western populations tend to be subtly darker than eastern ones. Juveniles are similar to adults but have faint barring on the chest.

In western and northeastern North America, the Brown Creeper occurs year-round. Elsewhere, it is present in spring and summer in the north, but migrates south for the winter. It feeds in a distinctive manner, climbing tree trunks like a mouse, using its tail as a support. Typically it spirals up a tree trunk, then drops to the base of an adjacent trunk to repeat the process.

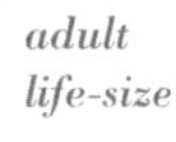

adult life-size

adult

FACT FILE

LENGTH 5.25 in (13.5 cm)

FOOD Invertebrates

HABITAT Wide range of forested habitats

STATUS Widespread and common resident and migrant

VOICE Song is a series of high-pitched *tsee-see-see* notes. Call is a thin *tsee*

Rock Wren

Salpinctes obsoletus

The Rock Wren is a dumpy-bodied songbird and an excellent songster. The sexes are similar. Adults have gray-brown upperparts, toned rufous on the rump and speckled white on the crown and back in particular. The tail is brown with dark barring and a pale tip. The face has speckled gray cheeks and a pale supercilium. The throat and chest are streaked grayish white and the underparts are otherwise pale, subtly suffused with orange-buff. Juveniles are similar to adults but with more uniform, less speckled upperparts.

In the north of its range, the Rock Wren is present as a breeding species mainly from May to September. Farther south it is present year-round, numbers boosted outside the breeding season as northern birds head south; the winter range extends to Mexico. There is also some altitudinal migration. Rock Wrens forage with a bouncing gait, and breeding birds create a pavement of flattened stones that leads to the nest entrance.

adult

adult
life-size

FACT FILE

LENGTH 6 in (15 cm)

FOOD Invertebrates

HABITAT Dry stony habitats with scree, boulders, and pebbles

STATUS Locally common summer visitor and partial resident

VOICE Song comprises various trilling whistles, many phrases repeated several times in succession, and a rattle. Calls include a shrill *ch-tsee*

Canyon Wren

Catherpes mexicanus

adult

The Canyon Wren is a distinctive long-billed bird of typically inaccessible terrain. The sexes are similar. Adults have reddish-brown upperparts with dark barring on the wings and tail, and tiny black and white spots on the back and rump. The crown is gray with dark streaks. The face, throat, and chest are clean white with a clear separation from the streaked and spotted rufous underparts. Juveniles are similar to adults but with less distinct spots and bars.

The Canyon Wren is present year-round in its favored canyon habitats in western North America. Although its appearance is distinctive, it feeds in an unobtrusive manner, making it a challenge to spot. Listen for its song or call to be certain it is present in potentially suitable habitat.

FACT FILE

LENGTH 5.8–6 in (14.5–15 cm)

FOOD Invertebrates

HABITAT Arid cliffs and canyons

STATUS Locally common resident

VOICE Song is a descending series of whistles ending with three or four soft, rasping notes. Call is a loud *jeet*

adult life-size

House Wren

Troglodytes aedon

The House Wren is a dumpy-bodied songbird with a needle-like bill. The tail, which is proportionately much longer than in Winter (p.111) and Pacific (p.110) wrens, is often cocked up. Subtle subspecies variation occurs across the species' range. Given this variation, the sexes are similar. Overall, all adults have brown plumage, darkest on the upperparts and with barring on the wings and tail. The face and throat range from buff in eastern birds to gray-buff in western birds. The underparts are otherwise pale with a variable buff suffusion to the breast and flanks, the latter also variably barred dark. On average, eastern birds are a richer brown than western birds, which tend to be grayer overall. Juveniles are similar to their adult counterparts but have a scaly look to the paler face and throat.

western race

adult life-size

The House Wren is present as a breeding species mainly from May to August across the north of its range. At other times of the year, it migrates south and occurs from southern U.S.A. to Mexico. Where summer and winter ranges overlap, the species is present year-round. It nests in treeholes and outbuildings.

FACT FILE

LENGTH 4.75 in (12 cm)

FOOD Invertebrates

HABITAT Wide range of habitats, including woodland, scrub, and gardens

STATUS Widespread and common summer visitor, local winter visitor, and very local resident

VOICE Song is an accelerating series of raspy trills, ending in a flourish. Call is a rasping *tche*

adult eastern race

Pacific Wren

Troglodytes pacificus

adult
life-size

The Pacific Wren is very similar to the Winter Wren (p.111), and formerly the two were considered to be conspecific. The sexes are similar. Adults have reddish-brown upperparts with barring on the wings and the very short tail, which is often held cocked up. The head is marked with a buff supercilium, and the underparts are warm buffish brown with dark barring on the flanks. The bill is needle-like and the legs are reddish. Juveniles are similar to adults but with subtly less distinct barring.

The Pacific Wren is present year-round across much of its range, which extends down the length of the Pacific coast, from mainland Alaska and its islands to southern California. Populations that breed inland in the northern Rockies move south in winter. The species forages for insects in deep undergrowth and its presence is often first detected by hearing its call or song.

FACT FILE

LENGTH 4 in (10 cm)

FOOD Invertebrates

HABITAT Mainly old-growth conifer forests

STATUS Common resident in much of its range; summer visitor to the interior

VOICE Song is variable and warbling, often ending in a trill. Call is a sharp *chip-chip*

adult

Winter Wren

Troglodytes hiemalis

adult

The Winter Wren is a dumpy, short-tailed songbird and very similar to the Pacific Wren (p.110); formerly the two were considered to be conspecific. The sexes are similar. Adults have reddish-brown upperparts with barring on the wings and the very short tail, which is often held cocked up. The head is marked with a pale buff supercilium, and the underparts are gray-buff with dark barring on the flanks. The bill is needle-like and the legs are reddish. Juveniles are similar to adults but with less distinct barring.

The Winter Wren is present as a breeding species across the north of its range, mainly from April to August. Outside the breeding season, it is found mainly in southeast U.S.A., although the species occurs year-round in parts of the northeast. It is often hidden from view as it forages in dense cover, creeping along like a mouse. Birds are sometimes seen briefly in flight, on whirring wingbeats.

FACT FILE

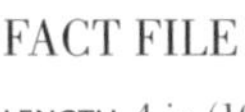

LENGTH 4 in (10 cm)

FOOD Invertebrates

HABITAT Dense woodland undergrowth and scrub

STATUS Widespread summer visitor, winter visitor, and local resident in southeast U.S.A.; present year-round in parts of the northeast

VOICE Song is variable and warbling, often ending in a trill. Call is a sharp *chip-chip*

adult life-size

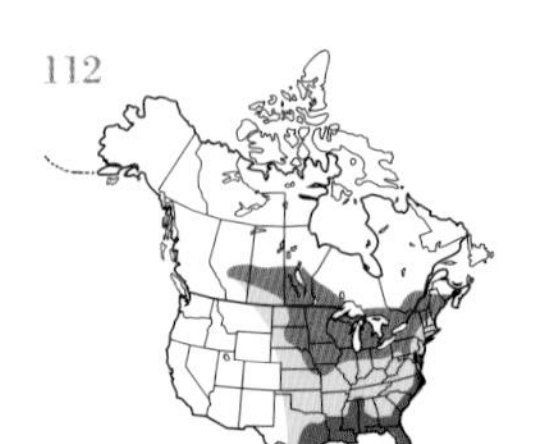

Sedge Wren

Cistothorus platensis

Although similar to a Marsh Wren (p.113), a Sedge Wren can be identified using plumage features, voice, and habitat. The sexes are similar. Adults have brown upperparts that are extensively but finely streaked on the crown and back, with dark barring on the tail. The underparts are buff, palest and whitish on the throat. The rufous wings are strongly barred and the head has a pale supercilium. Juveniles are similar to adults but less colorful.

adult

The Sedge Wren is a breeding visitor to central northern North America, where it is present mainly from May to September. After breeding, it migrates south and at other times of the year it is found in southeast U.S.A., where it occurs in coastal marshes as well as inland freshwater wetlands. It is generally unobtrusive and its presence is easiest to detect by listening for its call or song.

adult life-size

FACT FILE

LENGTH 4.5 in (11.5 cm)

FOOD Invertebrates

HABITAT Marshes and wet meadows where sedges flourish

STATUS Locally common migrant in both summer and winter

VOICE Song comprises a few sharp *chip, chip* notes followed by a dry rattle. Call is a sharp *chip*

Marsh Wren

Cistothorus palustris

adult

The Marsh Wren is a richly colorful songbird. The sexes are similar. Adults have reddish-brown upperparts with a few bold white streaks on the back (fewer but bolder than in the Sedge Wren; p.112). The underparts are pale but with a buffish-orange suffusion on the flanks. The wings and tail show subtle barring. The cheeks are buffish brown with white speckling, and there is a pale supercilium. Juveniles are similar to adults but are less strikingly marked and less colorful.

The Marsh Wren is present as a breeding species across much of northern central North America, mainly from May to August. It is present year-round in a few locations (generally coastal), but most birds move south to southern U.S.A. and Mexico for the winter. The species is rather secretive and a challenge to observe. Fortunately, it is vocal and its presence can be detected by listening for its call and song.

adult life-size

FACT FILE

LENGTH 5 in (12.5 cm)

FOOD Invertebrates

HABITAT Cattail (*Typha* spp.) marshes

STATUS Locally common migrant in both summer and winter

VOICE Song is variable but typically includes fluty, warbling notes; the tone is more liquid than the rasping notes of Sedge Wren (p.112). Calls include a sharp *tchut*

Carolina Wren

Thryothorus ludovicianus

The Carolina Wren is a well-marked and familiar little songbird. The sexes are similar, as are adults and juveniles. All birds have unstreaked, rich brown upperparts and warm buff underparts that are palest on the throat and breast. There is subtle barring on the wings and tail. The wing coverts have white tips that form incomplete wingbars (not present in the similar Bewick's Wren; p.115). The head is marked with a striking white supercilium and a speckled gray face. The bill is thin and downcurved, and the legs are reddish.

adult

The Carolina Wren is present year-round in its eastern North American range. It is mainly sedentary and so northern populations often crash in severe winters. The species is a familiar garden bird and by wren standards it is quite bold and easy to see. It could be confused with Bewick's Wren where their ranges overlap, but Carolina has warmer-looking brown plumage, incomplete white wingbars, and lacks the white tips to the tail feather of Bewick's.

FACT FILE

LENGTH 5.5 in (14 cm)

FOOD Invertebrates

HABITAT Woodlands, scrub, and gardens

STATUS Widespread and common resident

VOICE Song is a series of repeated fluty, whistling phrases. Call is a harsh, agitated *tchee-tchee-tchee*…

adult life-size

Bewick's Wren

Thryomanes bewickii

adult life-size

eastern race

Bewick's Wren is a long-tailed wren with an obvious pale supercilium. Regional plumage variation exists across its range, but given this variation the sexes are similar. Adults and juveniles have upperparts that are brown in eastern birds but grayer in western populations. The upperparts are unmarked except on the tail, which is barred and has white feather tips. The underparts are grayish white, palest on the throat, and with a rufous wash on the rear of the flanks in eastern populations.

Most Bewick's Wrens are present year-round in western North America, although birds in the far northeast of the species' range do tend to move south in winter. The long tail is often held cocked up and is sometimes flicked from side to side.

FACT FILE

LENGTH 5.25 in (13.5 cm)

FOOD Invertebrates

HABITAT Woodland, scrub, and gardens

STATUS Widespread and fairly common resident

VOICE Song is variable but usually comprises a series of wheezy phrases and ends in a trill. Calls include various harsh, rasping notes

adult western race

Cactus Wren

Campylorhynchus brunneicapillus

With its long tail and downcurved bill, this large desert wren looks more like a tiny thrasher than a true wren. The sexes are similar. Adults have brown or gray-brown upperparts with streaks on the back and dark barring on the wings and tail; the latter has white feather tips. The pale underparts are heavily spotted on the throat and breast; the belly and flanks are pale with dark spots in coastal populations but flushed orange-buff in inland birds. The head has a brown crown, striking white supercilium, and streaked gray-brown cheeks. Juveniles are similar to adults but less boldly marked.

The Cactus Wren is a locally common year-round resident of southwestern deserts. As its name suggests, it is often associated with cacti and will use them as a lookout perch or forage at their base for food.

adult

FACT FILE

LENGTH 8.5 in (21.5 cm)

FOOD Invertebrates

HABITAT Desert habitats

STATUS Locally common resident

VOICE Song is a rapid series of dry, chattering notes. Calls include various harsh notes

adult

adult
life-size

Blue-gray Gnatcatcher

Polioptila caerulea

Were it not for its long tail, the distinctive Blue-gray Gnatcatcher might be mistaken for a warbler. The sexes are dissimilar. Adult males have blue-gray upperparts and pale gray underparts. From above, the tail is black but with white outer feathers; from below, it is mostly white. The blackish wings have pale margins to the tertials and coverts. The head has a white eyering; the forehead is blackish in the breeding season but blue-gray at other times. Adult females are similar to a non-breeding male. Juveniles are similar to an adult female but sometimes have a subtle buff suffusion on the back.

The Blue-gray Gnatcatcher is present as a breeding species across temperate North America, mainly from April to August. It occurs year-round in parts of southern U.S.A. (particularly the southeast) and its winter range extends from there to Central America. Its active foraging behavior recalls that of a warbler. The long tail, often cocked up, aids certain identification.

female

male
life-size

FACT FILE

LENGTH 4.3 in (11 cm)

FOOD Invertebrates

HABITAT Deciduous woodland

STATUS Widespread and common summer visitor

VOICE Song is a series of thin notes, and sometimes includes mimicry of nearby songsters. Call is a soft, mewing *pweoe*

Black-tailed Gnatcatcher

Polioptila melanura

Plumage details, habitat preferences, and distribution help separate this species from the Blue-gray Gnatcatcher (p.118). The sexes are dissimilar. Adult males have a blue-gray back and, in the breeding season, a black cap; at other times of the year the cap is blue-gray. The white eyering is striking and the blackish wings have pale feather margins. From above and below, the tail is mainly black but with white tips to the outer feathers. Adult females are similar to a non-breeding male, and juveniles are similar to an adult female.

male

The Black-tailed Gnatcatcher is present year-round in its favored southwestern desert habitats. Look for it in areas of thorn scrub, where typically it is rather bold and easy to observe.

male
life-size

FACT FILE

LENGTH 4.3 in (11 cm)

FOOD Invertebrates

HABITAT Desert habitats

STATUS Locally common resident

VOICE Song comprises a series of phrases that are similar to the calls; these include rasping *tch-tch* and harsh *tchee-tchee* notes

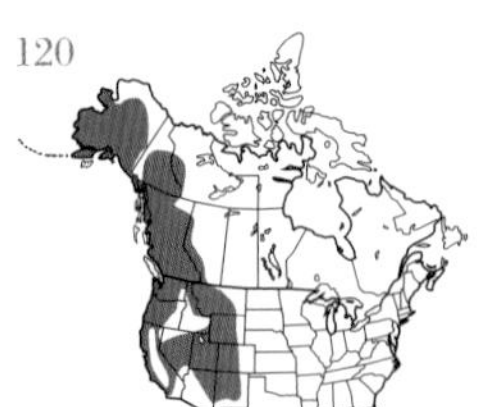

American Dipper

Cinclus mexicanus

The American Dipper is a dumpy-bodied bird that is always found near flowing water. The sexes are similar. Adults have dark blue-gray plumage with a brownish wash to the head that is visible in good light. When the bird blinks, it reveals white eyelids. The bill is dark and relatively stout, and the legs are pale. Juveniles are similar to adults but the wing feathers have pale margins and the bill is dull yellow.

adult

FACT FILE

LENGTH 7.5 in (19 cm)

FOOD Aquatic invertebrates

HABITAT Fast-flowing rocky streams and rivers

STATUS Widespread and locally common

VOICE Song comprises a prolonged series of whistled phrases, each repeated a few times. Call is a sharp *tzeet*

The American Dipper is present year-round within its western North American range. Most birds are sedentary, but those that breed in the mountains often have to undertake altitudinal migrations if waters freeze in winter. It is usually seen perched on boulders midstream, often bobbing up and down, or flying along a watercourse on whirring wingbeats. American Dippers dive and submerge to feed.

Golden-crowned Kinglet

Regulus satrapa

The Golden-crowned Kinglet is a tiny warbler-like songbird. A bold head pattern allows separation from the otherwise similar Ruby-crowned Kinglet (p.123). The sexes are dissimilar. Adult males have a gray-green back, and dark wings with two white wingbars. The head pattern comprises a white supercilium emphasized below by a dark eye stripe and above by a black margin to the golden-centered crown. The cheeks are gray and the throat and rest of the underparts are pale gray-buff. The legs are black and the feet are yellowish. Adult females and juveniles are similar to the male but the crown center is yellow.

The Golden-crowned Kinglet is present in its northern breeding range mainly from April to September. It occurs year-round in the northeast and parts of western North America, but most northern birds migrate south in fall, occurring throughout the southern half of the continent. The species is constantly active in search of food and sometimes joins roving mixed-species songbird flocks outside the breeding season.

female

male
life-size

FACT FILE

LENGTH 4 in (10 cm)

FOOD Invertebrates

HABITAT Northern conifer forests in summer; conifer and mixed forests in winter

STATUS Widespread and common, in northern latitudes in summer, and in the southern half of North America in winter

VOICE Song is a sweet *tswi, tswi, tswi, tswit-tswit-tswit*. Call is a thin *twsi*

Ruby-crowned Kinglet

Regulus calendula

female

Although this species is slightly bigger than the Golden-crowned Kinglet (p.122), the best identification feature for separating the two is its plain face. The sexes are dissimilar. Adult males have mainly gray-green upperparts, and dark wings with two white wingbars. The central ruby-colored crown patch is revealed only in displaying birds and is otherwise hidden. On the face, a pale patch surrounds and emphasizes the dark beady eye. The underparts are pale olive-gray. Adult females and juveniles are similar to the male but lack the ruby crown patch.

The Ruby-crowned Kinglet is present as a breeding species across northern latitudes of North America and western mountain ranges, mainly from May to September. Although present year-round locally in the southwest, most birds migrate south in fall and the winter range extends from southern U.S.A. to Central America. The species is extremely active and always on the go, searching for invertebrates among the foliage of trees and shrubs.

male life-size

FACT FILE

LENGTH 4.25 in (11 cm)

FOOD Invertebrates

HABITAT Northern conifer forests in summer; wide range of wooded habitats in winter

STATUS Widespread and common, with distinct summer and winter ranges

VOICE Song comprises thin *tsi-tsi-tsi* notes followed by a chattering warble. Call is a rasping disyllabic *ti-dit*

Eastern Bluebird

Sialia sialis

The Eastern Bluebird is a colorful and familiar songbird. The sexes are dissimilar. Adult males have blue upperparts, and a capped appearance to the head created by the orange-red partial collar. The throat, breast, and flanks are also orange-red, and the belly and undertail are white. Adult females have gray-brown upperparts with a blue tail and flight feathers. The throat, partial collar, and breast are suffused orange while the belly and undertail are white. Juveniles are brown, darker above than below, and with pale spots on the upperparts and scaly-looking underparts.

juvenile

The Eastern Bluebird is present as a breeding species in the north of its range mainly from April to September. It is present year-round in southern U.S.A. and its winter range extends to Mexico. The species nests in treeholes and will use nestboxes; it has suffered from competition for nest sites from European Starlings (p.164) and House Sparrows (p.348). Outside the breeding season it forms flocks, when berries and fruits feature in the diet.

female

male
life-size

FACT FILE

LENGTH 7 in (18 cm)

FOOD Invertebrates and fruit

HABITAT Open woodland, parks, and gardens

STATUS Widespread, fairly common but declining summer visitor; present year-round in the south

VOICE Song is a series of warbling phrases. Call is a disyllabic *tchu-lee*

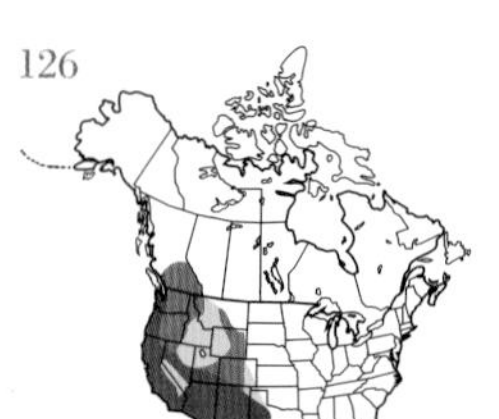

Western Bluebird

Sialia mexicana

The Western Bluebird is a colorful songbird. The sexes are dissimilar. Adult males have blue upperparts, including the head and throat, with a well-defined separation from the orange-red breast and flanks. The otherwise white underparts are suffused blue in the center of the belly. Adult females resemble a very dull male with gray-brown upperparts, dark wings and tail with a hint of blue, an orange-red breast, and otherwise pale gray underparts. Juveniles are brown, darker above than below, and with pale spots on the upperparts and scaly-looking underparts.

The Western Bluebird is present as a breeding species in the north of its range, mainly from April to August. Farther south, it is present year-round and its winter range extends into Mexico. It perches on bare branches and wires, and nests in treeholes as well as nestboxes; it suffers from competition for nest sites with European Starlings (p.164) and House Sparrows (p.348).

female

male

FACT FILE

LENGTH 7 in (18 cm)

FOOD Invertebrates and fruit

HABITAT Open woodland

STATUS Scarce summer visitor and local year-round resident

VOICE Song (sung at dawn) comprises a series of call notes, such as *chuu* and a disyllabic *chut-et*

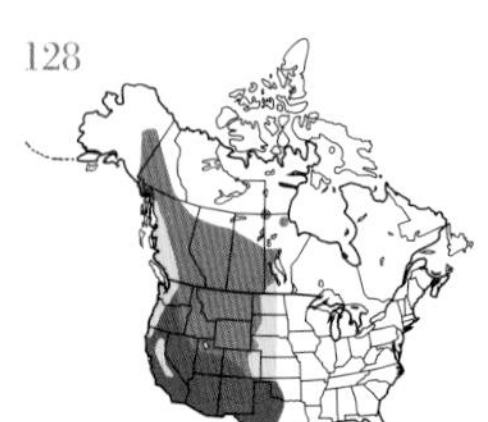

Mountain Bluebird

Sialia currucoides

A male Mountain Bluebird is a beautiful, relatively long-winged songbird. The sexes are dissimilar. Adult males have blue plumage, darkest on the wings and tail, and palest on the belly and undertail. Adult females have mainly gray upperparts, with blue on the wings and tail. The underparts are gray-buff with a variable orange-buff suffusion on the breast. Juveniles are similar to an adult female but the underparts have pale spots.

The Mountain Bluebird is present as a breeding species across much of its North American range from April to August. Outside the breeding season it moves south and to lower altitudes, and its winter range is mainly southwest U.S.A.; it is present year-round in parts of the region. Although they regularly perch on wires and branches, Mountain Bluebirds often hover while scanning for prey.

female

FACT FILE

LENGTH 7.25 in (18.5 cm)

FOOD Invertebrates and fruit

HABITAT Lightly wooded open habitats

STATUS Widespread and fairly common summer visitor; winters farther south, and present year-round in zone where seasonal ranges overlap

VOICE Song comprises a series of call-like notes such as *tchi-uu*

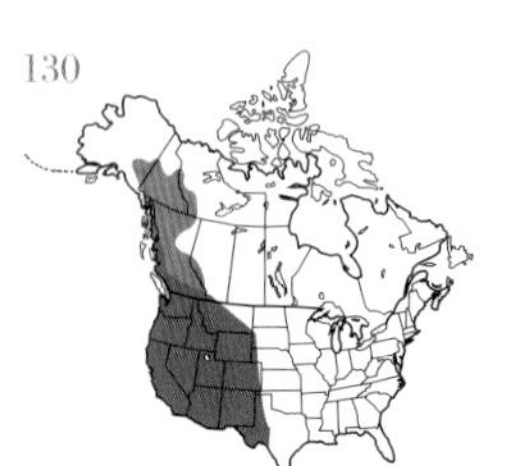

Townsend's Solitaire

Myadestes townsendi

Townsend's Solitaire is a slim, long-tailed songbird. The sexes are similar. Adults have mainly gray body plumage, darkest on the back and palest on the underparts. The head shows a white eyering, and the wings are dark with orange-buff patches and whitish tips to the tertials and greater coverts. The tail is black, with white edges and broad tips to the outer feathers. In flight, note the dark bar on the otherwise pale underwing coverts. Juveniles are brown with numerous pale spots, but with the same wing and tail markings as adults.

adult life-size

Townsend's Solitaire is present as a breeding species in the north of its range, mainly from May to August. Northern birds migrate south in fall and their range extends to southern mountain ranges, where the species is a year-round resident. The song and call are evocative of rugged, forested wilderness. The diet comprises mainly invertebrates in summer; berries, notably those of junipers (*Juniperus* spp.), are important in winter.

adult

adult

FACT FILE

LENGTH 8.5 in (21.5 cm)

FOOD Invertebrates and berries

HABITAT Rugged conifer forests

STATUS Widespread and common summer visitor; present year-round in the south of its range

VOICE Song comprises rich, warbling phrases. Call is a whistled *tiu, tiu…*, resembling a sonar blip

Veery

Catharus fuscescens

The Veery is a small, secretive thrush. The sexes are similar. Adults have reddish-buff upperparts. Markings on the gray-buff face are faint, and the pale buff throat is defined by a brown line. The breast is faintly suffused with yellow-buff and marked with subtle brown spots. The rest of the underparts are pale gray, with faint gray spots on the lower breast and flanks. Juveniles are brown and spotted, but by fall, when they migrate, their plumage has become similar to that of an adult aside from buff tips to the wing coverts.

adult

adult

The Veery is present as a breeding species, mainly from May to August. The rest of the year is spent in South America. The species is often associated with wetland willow scrub where it forages in the leaf litter for invertebrates. It can be a challenge to observe well.

FACT FILE

LENGTH 7.3 in (18.5 cm)

FOOD Invertebrates

HABITAT Wet deciduous woodland

STATUS Widespread and common summer visitor

VOICE Song is a fluty *vee, vedidi, vedidi, veer, veer*, each phrase descending in tone. Call is a sharp *veer*

adult life-size

Gray-cheeked Thrush

Catharus minimus

The Gray-cheeked Thrush is a small and fairly distinctive thrush. The sexes are similar. Adults have a gray-brown tail and upperparts. The face has gray cheeks (buff in the similar Swainson's Thrush; p.136) with a faint, pale gray eyering (Swainson's has buff "spectacles"). The pale throat and malar stripe are separated by a dark line. The breast is suffused with buffish yellow and adorned with dark spots; the otherwise pale underparts show faint gray spots on the lower breast and gray-washed flanks. Juveniles are gray-brown and spotted, but by fall, when they migrate, their plumage is similar to that of an adult aside from pale tips to the wing coverts.

adult

immature

The Gray-cheeked Thrush is present as a breeding species mainly from May to August. It spends the rest of the year in northern South America. It favors damp, boggy woodland and forages in leaf litter for invertebrates. It is generally unobtrusive and easily overlooked.

FACT FILE

LENGTH 7.25 in (18.5 cm)

FOOD Invertebrates

HABITAT Northern mixed and coniferous forests

STATUS Widespread and fairly common summer visitor

VOICE Song is a series of fluty notes, descending in tone, sometimes ending with a trill. Call is a nasal *piuur*

adult
life-size

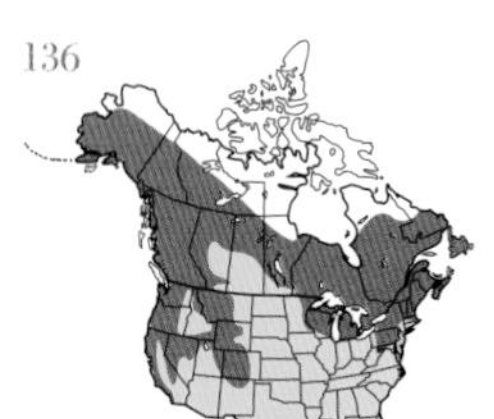

Swainson's Thrush

Catharus ustulatus

Swainson's Thrush is a small, subtly marked songbird. The sexes are similar. Adults have an olive-buff tail and upperparts, "colder" looking than the plumage of a Veery (p.132). The face and throat are buff, the latter bordered by a dark line, and the eye is framed by buff "spectacles." The underparts are pale, the breast having a buffish-yellow suffusion and dark spots; there are subtle gray spots on the lower breast and gray flanks. Juveniles are olive-brown and spotted, but by fall, when they migrate, their plumage resembles that of an adult aside from buff tips to the wing coverts.

adult life-size

FACT FILE

LENGTH 7 in (18 cm)

FOOD Invertebrates

HABITAT Conifer forests with a dense understory

STATUS Widespread and common summer visitor

VOICE Song comprises a series of fluty whistles that rise in tone throughout the sequence. Call is a sharp *quiirp*

Swainson's Thrush is present as a breeding species mainly from May to September. It passes the rest of the year in South America. It spends much of its time in dense cover, foraging for invertebrates among leaf litter. Its presence in an area is often first detected by hearing its song.

adult

adult

Hermit Thrush

Catharus guttatus

The Hermit Thrush is a fairly distinctive *Catharus* thrush. The sexes are similar. Adults have a dull brown cap and back that contrast with the rufous tail. The wings show a rufous panel on the primaries and the head has a white eyering. The pale throat is bordered by a black line. The underparts are pale, the breast suffused yellow and marked with black spots. Juveniles are brown and spotted, but by fall, when they migrate, their plumage resembles that of an adult aside from pale tips to the wing coverts.

adult life-size

FACT FILE

LENGTH 6.75 in (17 cm)

FOOD Invertebrates

HABITAT Conifer forests

STATUS Widespread and common summer visitor and local winter visitor

VOICE Song comprises a series of fluty whistles. Call is a soft *chuck-chuck*

The Hermit Thrush is present as a breeding species mainly from May to August. Most birds migrate south in fall, and its winter range extends from southern U.S.A. to Central America. Like its cousins, it forages for invertebrates among leaf litter. It raises its tail and flicks its wings, habits not seen in closely related thrush species.

American Robin

Turdus migratorius

The American Robin is an iconic songbird. The sexes are subtly dissimilar. Adult males have a gray-brown back, rump, and wings. The dark head has bold white "eyelids" framing the eye, and a dark-streaked white throat. The tail is dark brown, and the underparts are brick red with white on the lower belly and undertail. Reddish underwing coverts can be seen in flight. The legs and bill are yellowish. Adult females are similar to an adult male but much less colorful and with a hint of dark barring on the breast. Juvenile plumage recalls that of an adult female but with white spots on the back and dark spots on the underparts.

The American Robin is present as a migrant breeding species to the north of its range, mainly from April to September. It occurs year-round in the southern half of North America, where numbers are boosted in winter by migrants. It is a familiar sight in parks and gardens, and typically it is bold and easy to observe.

female

FACT FILE

LENGTH 10 in (25.5 cm)

FOOD Invertebrates, particularly earthworms, and berries

HABITAT Wide range of wooded habitats, parks, and gardens

STATUS Widespread and common summer visitor and year-round resident

VOICE Song comprises a series of whistling phrases, each often disyllabic, and separated by distinct pauses. Calls include a sharp *puup*. Flight call is a thin trill

male

male life-size

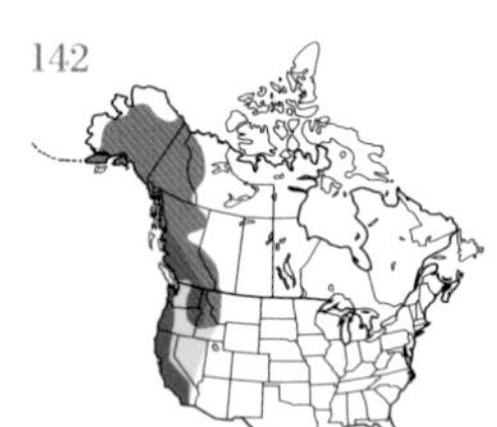

Varied Thrush

Ixoreus naevius

The Varied Thrush is a colorful and distinctive songbird. The sexes are dissimilar. Adult males have a blue-gray crown, back, and rump, and dark wings marked with orange-buff patches and wingbars. The face pattern comprises a black eye patch and orange-buff supercilium. A black breast band separates the throat from the rest of the orange-buff underparts, which show subtle bluish scaling on the flanks. Adult females are similar to an adult male, but brown replaces the bluish and black elements of the plumage. Juveniles are similar

female
life-size

to an adult female but with more scaly-looking underparts.

The Varied Thrush is present as a breeding species in the north of its range, mainly from May to September. Birds migrate south in fall and their winter range then overlaps in part with areas where the species is present year-round. In spring and early summer, the Varied Thrush's song is evocative of Pacific Northwest wilderness forests.

FACT FILE

LENGTH 9.5 in (24 cm)

FOOD Invertebrates and berries

HABITAT Damp conifer forests and willow scrub

STATUS Locally common summer visitor; resident or winter visitor elsewhere

VOICE Song is a series of well-spaced, strange-sounding whistles. Calls include thin whistles and a sharp *tchuup*

Wood Thrush

Hylocichla mustelina

The Wood Thrush is a plump-boded, colorful, well-marked songbird. The sexes are similar. Adults have rich reddish-brown upperparts. The face has pale, dark-streaked cheeks and a pale eyering. The throat is pale; the underparts are otherwise whitish but heavily marked with black spots. The legs are pink, and the bill is pink with a dark tip. Juveniles are brown and spotted, but by fall, when they migrate, their plumage is similar to that of an adult aside from pale tips to the wing coverts.

The Wood Thrush is present as a breeding species mainly from May to September. It spends the rest of the year in Central America. This secretive bird's song is much admired.

adult life-size

FACT FILE

LENGTH 7.5 in (19 cm)

FOOD Invertebrates and berries

HABITAT Deciduous woodland

STATUS Widespread and locally common summer visitor

VOICE Song comprises a series of fluty notes, and ends in a trill. Call is a rapid *ptt-ptt-ptt*

Wrentit

Chamaea fasciata

The Wrentit is a dumpy-bodied little songbird with a proportionately large head and very long tail. The plumage color varies subtly throughout the species' range, but in any given location the sexes are similar and adults and juveniles are alike. Northern adults have a brownish back and tail, grading to gray-brown on the crown and face. The eye has a pale iris and there is a subtle pale supercilium. The underparts are pinkish buff and streaked, with a gray-buff wash on the flanks. The legs and bill are dark. Southern adults are grayer overall than their northern counterparts, both above and below.

adult
life-size

FACT FILE

LENGTH 6.5 in (16.5 cm)

FOOD Invertebrates and berries

HABITAT Chaparral woodland

STATUS Locally common resident

VOICE Male's song is an accelerating series of whistling *pit* notes, ending in a trill; female's comprises evenly spaced notes and lacks a trill. Call is a churring rattle

The Wrentit is present year-round in its west coast range. Typically it favors rather dense oak woodland, and observing the species is not helped by its skulking and secretive habits. Once seen, however, it is hard to mistake for any other species.

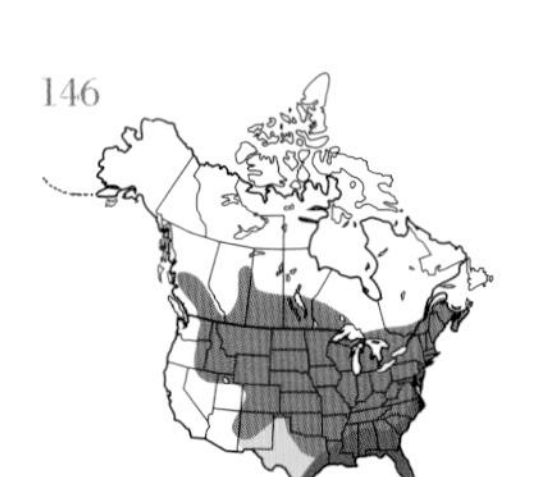

Gray Catbird

Dumetella carolinensis

adult

The Gray Catbird is a distinctive long-tailed songbird. The sexes are similar, and adults and juveniles are alike. All birds have deep blue-gray body plumage but with a black cap, and dark wings and tail. The brick-red undertail is diagnostic. The eye, legs, and bill are dark.

adult
life-size

The Gray Catbird is present as a breeding species mainly from May to August across much of eastern and north-central North America. In fall, birds migrate south and the species' winter range extends from coastal southeast U.S.A. to Mexico and the Caribbean generally. It is notoriously secretive and skulking, usually keeping to the cover of dense undergrowth, where it forages in leaf litter. The tail is often cocked up.

FACT FILE

LENGTH 8.5 in (21.5 cm)

FOOD Invertebrates and berries

HABITAT Dense woodland and scrub

STATUS Widespread and common summer visitor; local winter visitor

VOICE Song is a series of chattering squawks and whistles, often with some mimicry; the phrases are not repeated. Call is a cat-like meow

adult

Curve-billed Thrasher

Toxostoma curvirostre

The Curve-billed Thrasher is a long-tailed, curve-billed songbird of arid habitats. The sexes are similar. Adults have unmarked gray-brown upperparts. The wings have two pale wingbars and the tail has white feather tips. The eye has an orange iris and there is a faint pale supercilium. The pale throat is defined by a dark malar stripe, and the underparts are otherwise pale with faint brown spots that are boldest on the breast and flanks. Juveniles are similar to adults but the eye has a yellow iris, the bill is shorter, and the plumage markings are less intense.

The Curve-billed Thrasher is present year-round in its favored desert habitat. It is the most widespread and least skulking bird of its kind, sometimes seen perched on a cactus (particularly in spring) or searching for invertebrates on the ground.

juvenile

FACT FILE

LENGTH 11 in (28 cm)

FOOD Invertebrates and berries

HABITAT Deserts with plentiful cholla (*Cylindropuntia* spp.) and Saguaro (*Carnegiea gigantea*) cacti

STATUS Common resident

VOICE Song includes chirping trills and whistles, the phrases seldom repeated. Calls include a whistled *whit-Weet*

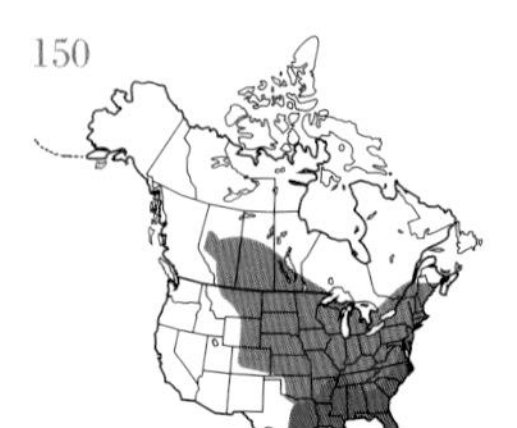

Brown Thrasher

Toxostoma rufum

The Brown Thrasher is a richly marked, long-tailed songbird with a slender, downcurved bill. The sexes are similar. Adults have reddish-brown upperparts, tail, and wings, the latter with two black and white wingbars. The face is streaked and the eye has a yellow iris. The underparts are buffish white with dark streaks on the breast, belly, and flanks. Juveniles are similar to adults but with dark eyes.

adult life-size

FACT FILE

LENGTH 11.5 in (29 cm)

FOOD Invertebrates, berries, and seeds

HABITAT Dense scrubby thickets

STATUS Widespread and common summer visitor; present year-round in the south of its range

VOICE Song is a series of fluty whistles, each phrase repeated two or three times. Calls include a sharp *stukk* and a softer *chrrr*

The Brown Thrasher is present as a breeding species across much of northern North America, mainly from May to August. Birds migrate south in the fall and the winter range extends across southeast U.S.A., with numbers boosting resident populations. The species is skulking and generally secretive, except in spring, when territorial males sing and become easier to locate.

adult

adult

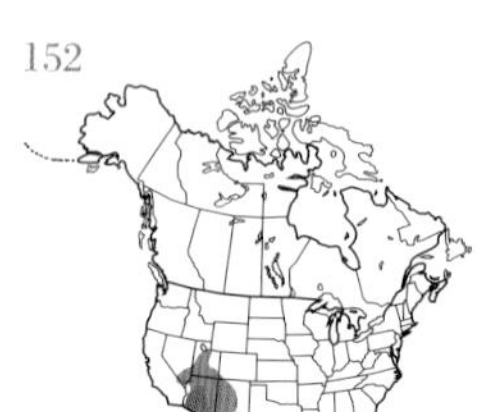

Bendire's Thrasher

Toxostoma bendirei

Bendire's Thrasher is a long-tailed desert songbird with a bill that is only slightly downcurved. The sexes are similar. Adults have unmarked sandy-brown upperparts. The wings have two indistinct buff wingbars and the dark tail has white feather tips. There is a faint pale supercilium and the eye has a yellowish iris. The pale throat is bordered by a buffish stripe. The otherwise mostly pale buff underparts are adorned with arrowhead spots, mainly on the breast and flanks, and the undertail is peachy buff. Juveniles are similar to adults but with less well-marked underparts.

adult life-size

FACT FILE

LENGTH 9.75 in (25 cm)

FOOD Invertebrates, berries, and seeds

HABITAT Desert habitats with scrub and grassland

STATUS Locally common resident

VOICE Song is a series of harsh warbling phrases, each repeated and well spaced. Call is a soft *tchuk*

Bendire's Thrasher is present year-round in yucca- and scrub-dominated desert habitats. From May to August there is potential for confusion with a juvenile Curve-billed Thrasher (p.148) where the species' ranges overlap; their calls and habitat preferences help with identification. Bendire's Thrasher is very secretive and the best chances of seeing it come in late winter and early spring, when males sing from open perches.

California Thrasher

Toxostoma redivivum

The California Thrasher is similar to its cousin the Crissal Thrasher (p.158), but their ranges do not overlap, they favor different habitats, and subtle structural differences also exist. In California Thrasher, the sexes are similar, as are adults and juveniles. All birds have unmarked, rich brown upperparts with brown underparts flushed orange-buff on the belly, flanks, and undertail. The pattern on the face comprises a faint buff supercilium and a pale throat bordered by a dark malar stripe. The eye is dark (Crissal has a yellow iris) and the bill, while long and downcurved, is marginally shorter than that of Crissal.

California Thrasher is present year-round in suitable chaparral woodland habitats in its narrow coastal Californian range. It is a secretive songbird, but once in a while it can be glimpsed running at speed from one patch of cover to another, tail cocked up. It is easiest to see in late winter and early spring, when males sing from prominent perches.

adult

adult

adult life-size

FACT FILE

LENGTH 12 in (30.5 cm)

FOOD Invertebrates, berries, and seeds

HABITAT Chaparral woodland and scrub

STATUS Locally common resident

VOICE Song often includes whistling and chattering phrases, with plenty of mimicry, each phrase repeated two or three times. Call is a soft *tchuk*

Le Conte's Thrasher

Toxostoma lecontei

Le Conte's Thrasher is a pale, long-tailed songbird with a strongly downcurved bill. The sexes are similar, as are adults and juveniles. All birds have unmarked, pale sandy-gray body plumage that is subtly paler below than above except for the orange-buff undertail. The tail itself is contrastingly dark, and the face pattern comprises a whitish throat defined by a fine dark malar stripe. The eye is dark.

adult

adult

Le Conte's Thrasher is present year-round in sparsely vegetated sandy deserts with scattered cholla cacti (*Cylindropuntia* spp.) and thorn bushes. It is much paler than the similar Crissal Thrasher (p.158) and favors very different habitats. Despite the open nature of its favored terrain, its secretive behavior means it is hard to see. Fortunately, in late winter males sing from prominent perches at dawn and dusk.

adult
life-size

FACT FILE

LENGTH 11 in (28 cm)

FOOD Invertebrates, berries, and seeds

HABITAT Open desert with sparse vegetation

STATUS Scarce and local resident

VOICE Song consists of a series of musical whistles, with some repetition of the phrases. Calls include a whistled *tsweep*

Crissal Thrasher

Toxostoma crissale

The Crissal Thrasher is similar in appearance to both California (p.154) and Le Conte's (p.156) thrashers, but its pale eye and habitat preferences help with identification. The sexes are similar, as are adults and juveniles. All birds have mostly unmarked gray-brown body plumage. The underparts are subtly paler than the upperparts, but the undertail is chestnut. The eye has a yellow iris, and the striking throat markings comprise a white throat, bordered by a black malar stripe and white "mustache" stripe.

The Crissal Thrasher is present year-round in its typically densely vegetated desert habitats. The presence of cover means this secretive bird is hard to see, except in early spring, when males sometimes sing from exposed perches. Feeding birds probe the ground with their long bills.

adult

FACT FILE

LENGTH 11.5 in (29 cm)

FOOD Invertebrates

HABITAT Scrubby desert habitats, often near washes and streams

STATUS Scarce resident

VOICE Song comprises musical whistles with some repetition of phrases or pairs of phrases. Call is a repeated *cheedalee, cheedalee…*

adult
life-size
adult

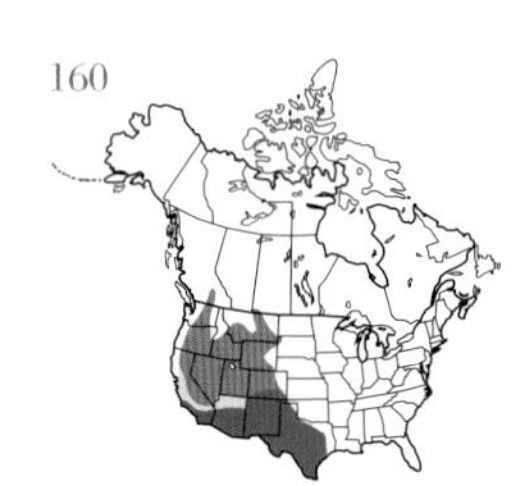

Sage Thrasher

Oreoscoptes montanus

Compared to other thrasher species, the Sage Thrasher has a relatively short bill and tail. The sexes are similar. Adults have faintly streaked gray-brown upperparts, and the wings show two white wingbars. The face pattern comprises a pale supercilium and pale-framed streaked cheeks. The eye has a yellow iris. The white throat is edged by a black malar stripe and the pale underparts are adorned with dark streaking everywhere except the undertail, which is unmarked orange-buff. Juveniles are similar to adults but with streaking on the upperparts and reduced streaking on the underparts.

The Sage Thrasher is present as a breeding species in upland sagebrush habitats mainly from March to September. Outside the breeding season, birds move south, and the winter range extends from southwest U.S.A. to Mexico. The species is less secretive than other thrashers and often perches on sagebrush bushes, making it relatively easy to see.

adult

FACT FILE

LENGTH 8.5 in (21.5 cm)

FOOD Invertebrates, berries, and seeds

HABITAT Sagebrush (*Artemisia* spp.) habitats

STATUS Locally common summer visitor; local winter visitor

VOICE Song comprises whistles and warbling phrases, with some repetition. Call is a sharp *tchuk*

adult
adult
life-size

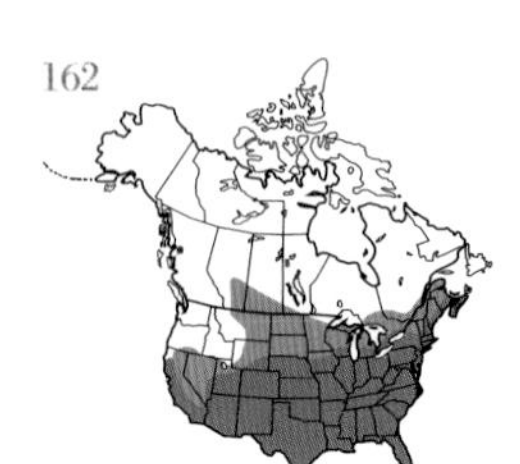

Northern Mockingbird

Mimus polyglottos

The Northern Mockingbird is a familiar bold songbird. The sexes are similar. Adults have gray upperparts and blackish wings with bold white wingbars and a white patch at the base of the primaries. The mainly black tail has white outer feathers. The yellow eye is emphasized by a dark eye stripe, and the bill is slightly downcurved. The underparts are pale gray, palest on the throat and undertail. Juveniles are similar to adults, but with pale buff underparts that are heavily spotted on the throat and breast.

The Northern Mockingbird is present year-round across much of southern U.S.A., although northern populations migrate south in fall. The species often perches prominently on wires and fenceposts, and often sings after dark in artificially lit suburbs.

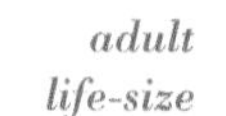

adult
life-size

FACT FILE

LENGTH 10 in (25.5 cm)

FOOD Invertebrates, fruits, and berries

HABITAT Wide range of wooded and lightly wooded habitats, including urban locations

STATUS Widespread and common resident, and partial summer migrant

VOICE Song consists of warbling phrases, with plenty of mimicry, each phrase repeated several times. Call is a sharp *tchek*

adult
adult

European Starling

Sturnus vulgaris

Although introduced little more than a century ago, the European Starling is now an abundant bird in North America. The sexes are subtly dissimilar in summer. Summer adult males have dark plumage that shows a green and purple iridescence. There is a blue base to the lower mandible of the otherwise yellow bill. Summer adult females are similar but have some pale spots on the underparts. In winter, all adults have dark plumage adorned with numerous white spots, and the bill is dark. Juveniles are gray-buff, palest on the throat, and have a dark bill; by winter, the body plumage has become dark with white spots, but the head and neck remain buff. The legs are reddish in all birds.

The European Starling is present year-round across much of North America, although northern populations move south in winter. It forms huge flocks in winter, which range widely in search of food.

winter

FACT FILE

LENGTH 9 in (23 cm)

FOOD Opportunistic omnivore

HABITAT Wide range of habitats, from farmland to towns and cities

STATUS Widespread and abundant resident, and partial migrant

VOICE Song includes various chatters, clicks, and whistles, plus mimicry of other birds and manmade sounds. Calls include chatters and drawn-out whistles

summer male
life-size

American Pipit

Anthus rubescens

winter

The American Pipit is a plump-bodied songbird that often bobs its tail as it walks. The sexes are similar. Summer adults have gray upperparts with subtle streaking on the back. The wings have two pale wingbars and pale margins to the tertials. The head has a pale supercilium and a pale unmarked throat. The underparts are otherwise buff, but with dark streaking on the breast and flanks. Winter adults are similar but have bolder streaking both on the back and on the underparts. Juveniles are even more boldly marked than a winter adult. The legs are dark in all birds.

The American Pipit is present in its breeding range mainly from May to August. Outside the breeding season birds migrate south and form large flocks in winter, favoring arable fields and short grassland.

FACT FILE

LENGTH 6.5 in (16.5 cm)

FOOD Mainly invertebrates, with some seeds

HABITAT Tundra and mountains in summer; open farmland in winter

STATUS Widespread and common summer visitor in the north; common winter visitor in the south

VOICE Song is a series of *tlee-tlee-tlee* notes, usually given in flight. Call is a thin *pi-peet*

summer adult
life-size

Sprague's Pipit

Anthus spragueii

adult

Sprague's Pipit is a secretive songbird and a prairie specialist. The sexes are similar. Adults have buffish-brown upperparts, heavily streaked on the back and crown. The wings show two white wingbars and the mainly dark tail has white outer feathers. The dark eye is emphasized by a pale buffish supercilium, cheeks, and lores. The underparts, including the throat, are pale, flushed buff on the flanks and breast, and with light streaking on the breast. The legs are pinkish. Juveniles are similar to adults but with bolder streaking.

Sprague's Pipit is present as a breeding species in its favored prairie grassland mainly from May to September; numbers are declining, its fate linked to the loss and degradation of its summer habitat. It winters from Texas south through Mexico. At all times it is a skulking species. It is easiest to see in spring, when males sing in flight.

adult
life-size

FACT FILE

LENGTH 6.5 in (16.5 cm)

FOOD Mainly invertebrates, but some seeds

HABITAT Prairie grassland in summer; rough grassland in winter

STATUS Rare summer migrant and winter visitor

VOICE Song is a descending series of whistled *tzee* notes, given in flight. Call is a thin *squeet*, repeated two or three times

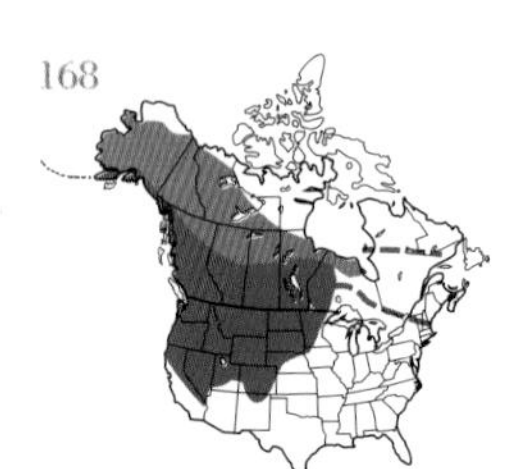

Bohemian Waxwing

Bombycilla garrulus

With its plump body and striking crest, the Bohemian Waxwing could only be confused with its cousin the Cedar Waxwing (p.170); subtle plumage differences allow separation. The sexes are subtly dissimilar. Adult males are mainly pinkish buff, palest and grayest on the belly. The head has a crest, a black mask, and a black throat. The dark primaries have white and yellow edges, the secondaries have red wax-like projections, and there is a white bar at the base of the primary coverts. The rump is gray, the undertail is chestnut, and the dark tail has a broad yellow tip. Adult females are similar to an adult male but with a less extensive black throat, and a narrower yellow tip to the tail. Juveniles are streaked buff; by winter their plumage is similar to an adult but they lack the red wax-like projections and pale margins to the primaries.

1st-winter female

immature

FACT FILE

LENGTH 8.25 in (21 cm)

FOOD Invertebrates in summer; berries in winter

HABITAT Boreal forests in summer; anywhere with berry-bearing bushes in winter

STATUS Widespread and common summer visitor; nomadic in winter, when it has an unpredictable range

VOICE Does not sing. Call is a vibrant trill

The Bohemian Waxwing is present in its northern summer breeding range from April to September. Outside the breeding season it forms roaming flocks that move south; the winter range extends well beyond the zone where birds are usually present year-round. In years when berries are in short supply, flocks move much farther afield.

male
life-size

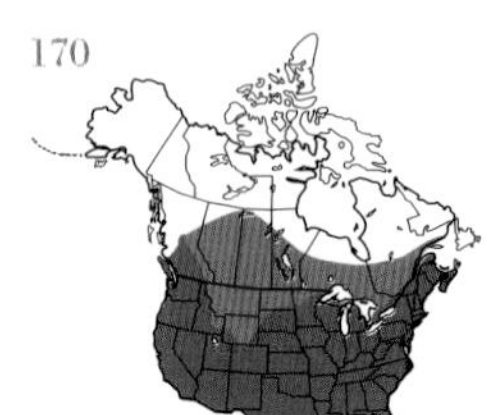

Cedar Waxwing

Bombycilla cedrorum

Superficially similar to the Bohemian Waxing (p.168), the Cedar Waxwing can be distinguished by subtle plumage differences. The sexes are similar. Adults have orange-buff plumage (pinkish buff in Bohemian), palest on the underparts and white on the undertail (chestnut in Bohemian). The dark wings have white inner margins to the tertials and red wax-like feather projections on the secondaries. The head has a crest, and a dark mask framed by white lines. The rump is gray and the dark tail has a yellow tip, subtly broader in males than in females. Juveniles are gray-buff; by winter they are similar to an adult but lack the red waxy wing projections.

The Cedar Waxwing is present in its northern summer breeding range mainly from May to September. Outside the breeding season birds move south and form roaming flocks. It is present year-round across much of central North America, but its winter range also extends south into Mexico. Winter flocks are nomadic, wandering in search of berry bushes.

FACT FILE

LENGTH 7.25 in (18.5 cm)

FOOD Invertebrates in summer; berries at other times

HABITAT Open woodland

STATUS Present year-round across much of central North America; common summer visitor farther north, and widespread in winter farther south

VOICE Does not sing. Calls include a buzzing *tzeee* note

Phainopepla

Phainopepla nitens

With its long tail and spiky crest, the Phainopepla is easily recognized. The sexes are dissimilar. Adult males have glossy black plumage and a red eye. A white wing patch is visible only in flight. Adult females are buffish gray overall, with a subtle blue tint to the head and white margins to the wing feathers. Juveniles are similar to an adult female but the plumage is overall buff and the eye is dark.

The Phainopepla's habitat requirements vary throughout the year. In the summer months, when its range extends northwards, it favors riverside woodland and thickets, while in winter its chosen habitat is desert scrub. It raises broods in both habitats.

FACT FILE

LENGTH 7.5–8 in (19–20 cm)

FOOD Invertebrates, fruits, and berries, particularly those of mistletoe (*Phoradendron* spp.)

HABITAT Riverine woodland in summer; desert habitats in winter

STATUS Locally common resident and summer visitor

VOICE Song includes whistles, warbles, and some mimicry. Call is a whistled *wu-ip*

male life-size

Olive Warbler

Peucedramus taeniatus

Despite its name and appearance, the Olive Warbler is not closely related to the wood warblers that belong to the family Parulidae. The sexes are subtly dissimilar. Adult males have a gray back and nape. The dark wings have two white wingbars, white-edged flight feathers, and a white patch at the base of the primaries. The tail is dark above with white outer feathers, and white below with dark outer tips. There is a dark mask, but the head, neck, and chest are otherwise reddish orange. The belly is gray and the undertail is white. Adult females are similar to an adult male but orange elements of the plumage are replaced by yellow. Immatures are similar to their respective adults but less colorful, and with an incomplete mask and less striking wing markings.

The Olive Warbler is a summer breeding visitor to the north of its range, present mainly from April to September. Present year-round farther south, its resident range extends into Mexico. Olive Warblers flick their wings as they forage.

male life-size

female

FACT FILE

LENGTH 5.25 in (13.5 cm)

FOOD Invertebrates

HABITAT Southern mountain conifer forests

STATUS Very locally common summer visitor and scarce year-round resident

VOICE Song is a trilling *tuet-tuet-tuet…* Call is a soft *tui*

Lapland Longspur

Calcarius lapponicus

female

The Lapland Longspur is a plump-bodied songbird that shuffles along the ground when feeding. The sexes are subtly dissimilar. Summer adult males have a streaked back, a chestnut nape, and a black crown, face, throat, and breast, with a white line running from the eye to the base of the wings. The wings have reddish-brown greater coverts and tertial edges. The belly and undertail are white but streaked black on the flanks. Winter adult males are similar, but black elements of the plumage are mostly replaced by streaked brown or buff; the throat is white. Adult females recall a winter male but are less colorful overall. Juveniles are similar to an adult female but with bolder streaking.

Lapland Longspurs breed in the Arctic mainly from April to August. They migrate south in fall, and the winter range extends across lowland southern U.S.A. Outside the breeding season the species forms flocks.

FACT FILE

LENGTH 6.25 in (16 cm)

FOOD Invertebrates in summer; mainly seeds in winter

HABITAT Tundra in summer; short grassland in winter

STATUS Widespread summer visitor in the north; locally common winter visitor to the south

VOICE Song comprises bursts of warbling phrases and whistles. Call is a rattle, typically given in flight

male life-size

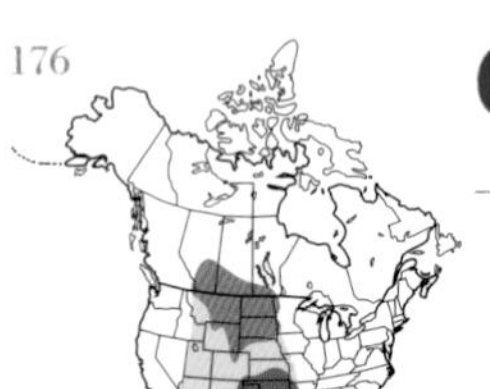

Chestnut-collared Longspur

Calcarius ornatus

The Chestnut-collared Longspur is a well-marked songbird, particularly in the breeding season. The sexes are dissimilar. Summer adult males have a streaked back, chestnut crown, black crown and ear covert margins, and a pale yellow-tinged face and throat. The wings have pale feather margins and a small white shoulder patch. The breast and belly are usually black, while the undertail is white. Winter adult males recall a breeding male but with muted colors and pale feather fringes that mask black elements of the plumage. Adult females recall a dull winter male and juveniles are duller still. All birds have a white tail with a central black tip.

This is a prairie-breeding species. Outside the breeding season it forms flocks and occurs from southern U.S.A. to Mexico.

female

male life-size

FACT FILE

LENGTH 6 in (15 cm)

FOOD Invertebrates in summer; mainly seeds in winter

HABITAT Prairie grassland in summer; dense grassland in winter

STATUS Local common summer visitor; local in winter

VOICE Song is a warbling *tsi tsididi tser-de*. Call is a rattle, given in flight

Smith's Longspur

Calcarius pictus

All Smith's Longspurs have orange-buff underparts but only the summer male is distinctive. The sexes are dissimilar. Adult summer males have a streaked brown back and brown wings with two white wingbars and a white "shoulder" patch. The head has striking markings, with a white supercilium, lores, and ear patch on an otherwise black cap. The neck, throat, and rest of the underparts are orange-buff. Birds in all other plumages are similar to a breeding male but have subdued colors and markings, the black elements of the head pattern being replaced by streaked brown, and white replaced by buff; the white "shoulder" patch is absent.

Smith's Longspur is present as a breeding species on northern tundra, mainly from June to August. Outside the breeding season it forms flocks and migrates south, wintering in a localized region of short grassland from Iowa to Texas. Although it lives in open habitats, it is hard to locate since it creeps around in cover and only reluctantly ventures out.

FACT FILE

LENGTH 6.25 in (16 cm)

FOOD Invertebrates in summer; seeds at other times

HABITAT Tundra in summer; grassland at other times

STATUS Widespread but scarce summer visitor; very local in winter

VOICE Song comprises whistling warbled phrases. Call is a dry rattle, given in flight

male
life-size

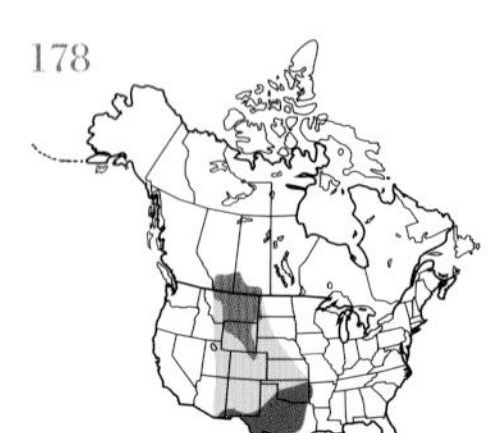

McCown's Longspur

Rhynchophanes mccownii

McCown's Longspur sexes are dissimilar. Summer adult males have a streaked buff back and a chestnut band (median coverts) on the wings. The head is palest on the throat and around the eye, and it has a black cap and malar stripe. There is a broad, dark breast band and the underparts are otherwise pale but with dark-flecked flanks. Adult males at other times of the year are similar, but black plumage elements are masked by pale feather fringes. Adult females are similar to a non-breeding male but have a brown breast; they are palest in winter. Juveniles are similar to a winter female, but with pale feather margins on the back and more obvious streaking on the underparts. All birds have a mostly white tail with a black inverted "T" (central feathers and feather tips).

McCown's Longspur breeds in northern prairie grassland mainly from April to August. Outside the breeding season it forms flocks and migrates south; its winter range extends from southern U.S.A. to northern Mexico. Because of habitat loss its population is in decline.

female

male life-size

FACT FILE

LENGTH 6 in (15 cm)

FOOD Invertebrates in summer; seeds at other times

HABITAT Shortgrass prairie in summer; short grassland in winter

STATUS Generally rather scarce summer visitor; local in winter

VOICE Song is series of chattering, warbling phrases, often delivered in flight. Call is a rattle, delivered in flight

Snow Bunting

Plectrophenax nivalis

immature

The sexes of this plump songbird are dissimilar. Summer adult males are mainly white but with a black back, outer flight feathers, leading edge of the wing, and tail center. The bill is black. Males in winter are similar, but the back feathers have orange-buff fringes and there is a similar color on parts of the face, wings, and flanks. The fringes wear during winter, revealing black and white plumage by spring. The bill is yellow. Adult females resemble a seasonal counterpart male, but white plumage elements are suffused orange-buff and black feathers are fringed brown. Juveniles are streaked brown but by winter they resemble winter adults, but with an orange-buff suffusion on the face and underparts. All birds show extensive white in the wings in flight.

The Snow Bunting is a tundra breeding species, mainly from May to September. At other times, it forms roaming flocks and migrates south. Its winter range extends across southern Canada and northern and central U.S.A.

FACT FILE

LENGTH 6.75 in (17 cm)

FOOD Invertebrates in summer; mainly seeds at other times

HABITAT Tundra in summer; short grassland in winter

STATUS Widespread and common summer visitor; widespread in winter but its precise range is unpredictable

VOICE Song is a series or warbling whistles. Calls include a soft *tiu*, often given in flight

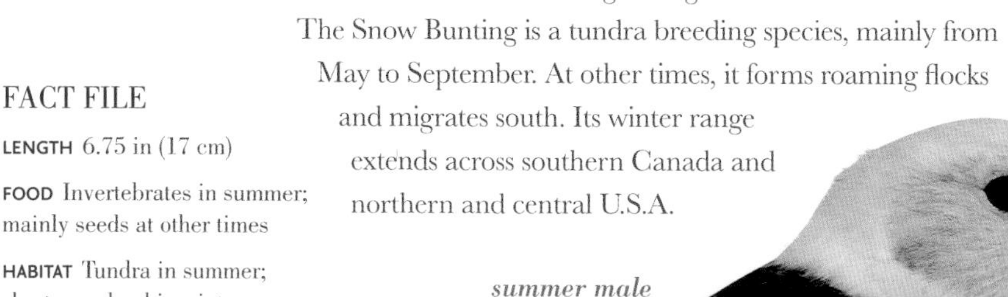

summer male life-size

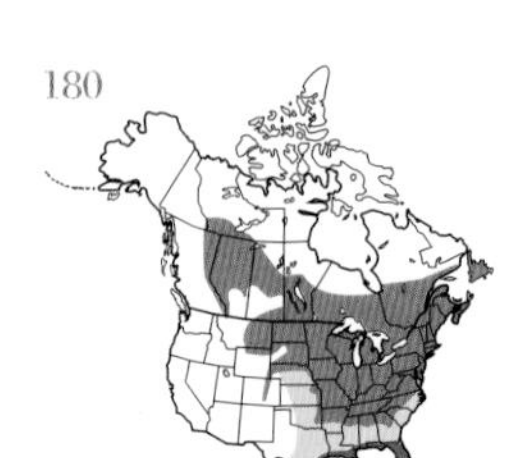

Ovenbird

Seiurus aurocapilla

The Ovenbird is a ground-dwelling wood warbler that looks a bit like a miniature thrush. The sexes are similar. Adults have mostly olive-brown upperparts and wings. The head is marked with a bold, black-bordered orange crown and a white eyering. The throat is white with a black malar stripe; the underparts are otherwise white with bold black spots and streaks on the breast and flanks. The legs are pinkish. Immatures are similar to adults but have two faint pale wingbars.

adult

The Ovenbird is present as a breeding species mainly from May to August. At other times of year it is found in Central America, although small numbers do overwinter in Florida and the very south of Texas. It is easily overlooked when foraging in leaf litter for invertebrates. Its presence is often first detected by hearing its song.

FACT FILE

LENGTH 6 in (15 cm)

FOOD Invertebrates

HABITAT Deciduous and mixed forests

STATUS Widespread and common summer visitor

VOICE Song is a whistled *ke-Chee, ke-Chee, ke-Chee*. Call is a sharp *tsik*

adult life-size

Worm-eating Warbler

Helmitheros vermivorum

adult

The Worm-eating Warbler has a striking head pattern but otherwise rather unremarkable plumage. The sexes are similar, as are adults and immatures. All birds have a buffish-brown back, wings, and tail. The head is overall yellow-buff but with a long black eye stripe and a long black line bordering the side of the crown. The underparts, including the throat, are warm buff, most intensely colored on the breast. The legs are pale pink and the bill is relatively long.

The Worm-eating Warbler is present as a breeding species mainly from May to August. It spends the rest of the year in Central America. It is unobtrusive in its habits, typically searching for caterpillars among bunches of hanging dead leaves.

FACT FILE

LENGTH 5.25 in (13.5 cm)

FOOD Invertebrates

HABITAT Deciduous and mixed woodland; often associated with steep slopes and cliffs

STATUS Locally common summer visitor

VOICE Song is a rapid trill. Call is a sharp *tsip*

adult life-size

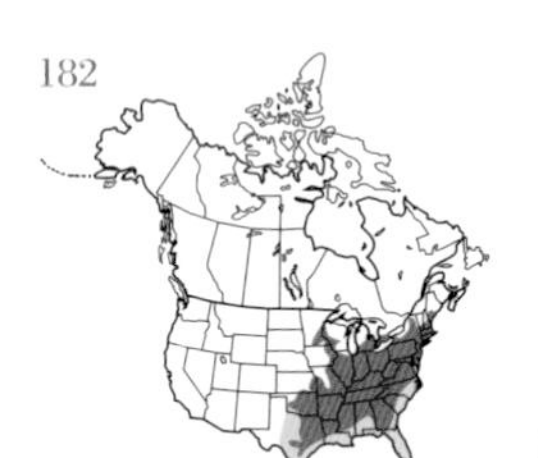

Louisiana Waterthrush

Parkesia motacilla

Similar to the Northern Waterthrush (p.183), the Louisiana Waterthrush is best separated from that species by studying its head markings and voice. The sexes are similar, as are adults and immatures. All birds have dark olive-brown upperparts, including the crown, back, wings, and tail. The head has a long, pale supercilium that is wider behind the eye than in front (even wider in Northern), and buffish in front of the eye but white behind (uniformly buff in Northern). The pale throat is unmarked (subtly streaked in Northern), and the otherwise pale underparts are suffused orange-buff on the rear of the flanks, with dark streaks on the breast, belly, and flanks. The legs are bright pink.

The Louisiana Waterthrush breeds mainly from April to August. Very small numbers winter in southeast U.S.A. but most are found in Central America outside the breeding season. This unobtrusive bird is easily overlooked.

adult

adult life-size

FACT FILE

LENGTH 6 in (15 cm)

FOOD Invertebrates

HABITAT Waterside and damp woodland

STATUS Widespread and common summer visitor

VOICE Song is a whistled *tseeu tsewit*. Call is harsh *tchtt*

Northern Waterthrush

Parkesia noveboracensis

The Northern Waterthrush is a well-marked ground-dwelling wood warbler. The sexes are similar, as are adults and immatures. All birds have dark olive-brown upperparts, including the crown, back, wings, and tail. The head has a narrow, pale supercilium that is uniformly wide and buffish along its length (supercilium of Louisiana Waterthrush (p.182) broadens behind the eye and is buff in front of the eye, but white behind). The pale underparts are suffused pale yellow, and marked with dark streaks on all areas except the undertail, and most intense on the breast and flanks (Louisiana Waterthrush has an unmarked throat). The legs are dull pink.

adult

The Northern Waterthrush is present as a breeding species, mainly from May to August. It spends the rest of the year in Central America and northern South America. As its name suggests, this ground-dwelling wood warbler is typically found near water, be that forest pools or stream margins.

adult
life-size

FACT FILE

LENGTH 5.75 in (14.5 cm)

FOOD Invertebrates

HABITAT Wet wooded habitats

STATUS Widespread and common summer visitor

VOICE Song is a rich *tuit-tuit-tuit, tchu-tchu-tchu*. Call is a thin *tzink*

Blue-winged Warbler

Vermivora cyanoptera

The Blue-winged Warbler is a colorful wood warbler. The sexes are dissimilar. Adult males have a mainly yellow head and underparts, and an olive-yellow nape and back. The head has a thin black eye stripe, and the bluish wings have two white wingbars. Adult females are less colorful than an adult male and have duller wingbars. Immatures are similar to, but duller than, an adult female.

Brewster's Warbler, male

The Blue-winged Warbler hybridizes with the Golden-winged Warbler (p.186), resulting in distinct hybrids known as Brewster's and Lawrence's warblers. A male Brewster's has a mostly pale gray head and upperparts, and whitish underparts; yellow is restricted to the forehead and breast. A male Lawrence's is like a dull male Blue-winged with a dark throat and eye patch. Hybrid females have duller plumage than their hybrid male counterparts.

Blue-winged Warbler, male life-size

The Blue-winged Warbler is present as a breeding species mainly from May to August. It spends the rest of the year in Central America and the Caribbean region. It searches unobtrusively for insects among the foliage of shrubs and small trees.

Blue-winged Warbler, female

Lawrence's Warbler, female

FACT FILE

LENGTH 4.75 in (12 cm)

FOOD Invertebrates

HABITAT Scrub-colonized meadows and young woodland

STATUS Widespread and locally common summer visitor

VOICE Song is typically a two-part buzzing trill, each phrase with different tones. Call is a sharp *tsik*

Golden-winged Warbler

Vermivora chrysoptera

female

The Golden-winged Warbler is a colorful little songbird. The sexes are dissimilar. Adult males have a blue-gray back, nape, and rear of crown, and a yellow forecrown. The blue-gray wings have a bright yellow panel. The head is marked with a black mask and throat, defined by a white supercilium and malar stripe. The pale underparts are suffused blue-gray on the flanks. Adult females are similar to males but black elements of the head pattern are replaced by streaked gray. Immatures are less colorful than their respective adults. All birds have a thin bill and black legs.

The Golden-winged Warbler is present as a breeding species mainly from May to August. It spends the rest of the year in Central America and the Caribbean region. It is an active feeder, searching for insects among the foliage of shrubs and trees. It hybridizes with Blue-winged Warbler (see p.184 for more details).

male
life-size

FACT FILE

LENGTH 4.75 in (12 cm)

FOOD Invertebrates

HABITAT Recently colonized wetland scrub

STATUS Widespread but generally scarce and declining summer visitor

VOICE Song is a high-pitched buzzing *tzee-dee-dee*. Call is a thin *tsip*

Black-and-white Warbler

Mniotilta varia

The Black-and-white Warbler certainly lives up to its common name. The sexes are subtly dissimilar. Adult summer males have striped black and white upperparts, and the black wings have two white wingbars. The throat is black and the underparts are otherwise mainly white but streaked black on the flanks. By fall, males have acquired a white throat and gray ear coverts. Adult summer females are like a fall male; by fall they sometimes acquire a subtle yellow suffusion to their underparts. Immatures are similar to their respective fall adults.

female

The Black-and-white Warbler is present as a breeding species mainly from May to August. It spends the rest of the year in Central America. Unlike most other wood warbler species, it often probes crevices in bark for insects with its slender bill, in the manner of a nuthatch or Brown Creeper (p.106).

summer male life-size

FACT FILE

LENGTH 5.25 in (13.5 cm)

FOOD Invertebrates

HABITAT Wide range of wooded habitats

STATUS Widespread and common summer visitor

VOICE Song is a thin, repeated *weesa, weesa*... Call is a sharp *tchak*

Prothonotary Warbler

Protonotaria citrea

female

The Prothonotary Warbler is a colorful and distinctive songbird with a relatively long, slender bill. The sexes are subtly dissimilar. Adult males have a bright yellow head, neck, and underparts. The back is olive-brown and the wings are blue-gray with pale feather margins. The mainly dark tail has bold white spots, and the legs and bill are dark. Adult females and immatures are similar to adult males but less colorful, and with smaller spots on the tail.

The Prothonotary Warbler is present as a breeding species mainly from May to August. It spends the rest of the year in coastal regions of Central America and northern South America, typically in mangrove forests. Unusually among wood warblers, it nests in treeholes and will use a nestbox.

male
life-size

FACT FILE

LENGTH 5.5 in (14 cm)

FOOD Invertebrates

HABITAT Damp woodland, often near rivers

STATUS Locally common summer visitor

VOICE Song is a series of liquid *swiit-swiit-swiit…* notes. Call is a sharp *tchip*

Swainson's Warbler

Limnothlypis swainsonii

Swainson's Warbler is a rather plain-looking little songbird. The sexes are similar, as are adults and immatures. All birds are gray-brown on the wings, back, and nape, and have a rufous crown. The face is overall rather pale but marked with a brown eye stripe and gray-brown cheeks; these frame the long, pale supercilium. The underparts, including the throat, are pale with a subtle gray suffusion on the flanks.

adult

Swainson's Warbler is present as a breeding species in southeast U.S.A. mainly from April to September. It spends the rest of the year in Central America and the Caribbean region. It is one of the trickiest wood warblers to observe, partly because it is secretive but also because it forages in deep undergrowth.

adult
life-size

FACT FILE

LENGTH 5.5 in (14 cm)

FOOD Invertebrates

HABITAT Inundated woodland in valley bottoms and damp rhododendron thickets

STATUS Scarce summer visitor

VOICE Song is a clear *twee-twee-twe-tsu-twe-tsu*. Call is a sharp *tchip*

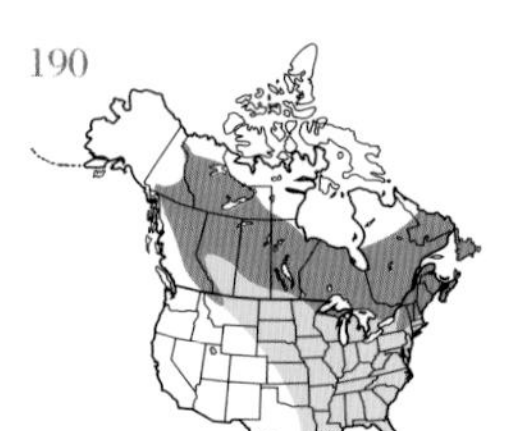

Tennessee Warbler

Oreothlypis peregrina

The Tennessee Warbler is an active little songbird. The sexes are dissimilar. Adult males have an olive-green back and an olive-brown tail and wings, the latter with indistinct pale wingbars. The head is blue-gray with a dark eye stripe and a whitish supercilium. The underparts, including the throat, are whitish with a gray suffusion on the flanks. Adult females are similar to a male but less colorful and with a yellowish suffusion on the neck and breast. Adults of both sexes are more colorful in spring and summer than in fall. Immatures are buffish yellow with pale wingbars and a pale supercilium. The underparts are mostly pale buff but whitish on the undertail. In all birds the bill and legs are dark.

The Tennessee Warbler is present as a breeding species across much of northern North America, mainly from May to July. It spends the rest of the year in Central and South America. The species is usually easy to see both on its breeding grounds and on migration.

female

male life-size

FACT FILE

LENGTH 4.75 in (12 cm)

FOOD Invertebrates, notably spruce budworms (larvae of moths belonging to the genus *Choristoneura*)

HABITAT Northern conifer forests

STATUS Widespread and common summer visitor

VOICE Song is a three-part series of *chip* notes, each part increasing in delivery speed. Call is a sharp *tscht*

Orange-crowned Warbler

Oreothlypis celata

The Orange-crowned Warbler's orange crown patch is hard to see and not a useful field character. The body plumage shows subtle geographical variation across its breeding range, with the west coast subspecies being brighter yellow than the duller yellow-green subspecies found across northern North America. The sexes are dissimilar. Given the regional variation, adult males are yellow-green, subtly darker above than below, with faint streaks on the underparts and a yellow-buff undertail (cf. immature Tennessee Warbler; p.190). The face shows a gray tint, an indistinct pale eyering, and a pale supercilium. Adult females and immatures are similar to males but grayer overall and less colorful.

The Orange-crowned Warbler is present as a breeding species mainly from May to September. It spends the rest of the year in Central America.

adult

adult
life-size

FACT FILE

LENGTH 5 in (12.5 cm)

FOOD Invertebrates

HABITAT Deciduous woodland

STATUS Widespread and common summer visitor

VOICE Song is a warbling trill whose pitch drops from start to finish. Call is a sharp *tsik*

Lucy's Warbler

Oreothlypis luciae

Lucy's Warbler is a small, pale songbird. The sexes are subtly dissimilar. Adult males have pale blue-gray upperparts except for a chestnut rump and dark gray wings and tail. The underparts are pale gray with a white undertail. The pale face emphasizes the dark eye and there is a chestnut crown that is often partly concealed. Adult females are similar to an adult male but with a much-reduced crown patch and paler face. Immatures are similar to an adult female but the crown patch is absent.

male life-size

Lucy's Warbler is present as a breeding species mainly from April to July. It spends the rest of the year on the west coast of Mexico. Unusually among wood warblers, it nests in treeholes and crevices.

FACT FILE

LENGTH 4.25 in (11 cm)

FOOD Invertebrates

HABITAT Waterside mesquite woodland

STATUS Locally common summer visitor

VOICE Song is a rapid warbling trill. Call is sharp *chink*

female life-size

Nashville Warbler

Oreothlypis ruficapilla

female

The Nashville Warbler is a colorful little songbird. The sexes are subtly dissimilar. Adult males have an olive-green back, with a dull brown tail and wings. The head and neck are blue-gray except for the small orange crown patch, a white eyering, and a yellow throat; the rest of the underparts are also bright yellow. The legs are dark. Adult females are similar to an adult male but less colorful and with a reduced crown patch. Immatures are similar to an adult female but even paler, with a whitish throat and belly and no crown patch.

male life-size

The Nashville Warbler is present as a breeding species mainly from May to August. It spends the rest of the year in Central America. It is often associated with the tangled vegetation of colonizing secondary-growth scrub and woodland.

FACT FILE

LENGTH 4.75 in (12 cm)

FOOD Invertebrates

HABITAT Deciduous and mixed forest

STATUS Widespread and common summer visitor

VOICE The two-part song comprises bouncy, whistling phrases followed by trills. Call is a thin *tsik*

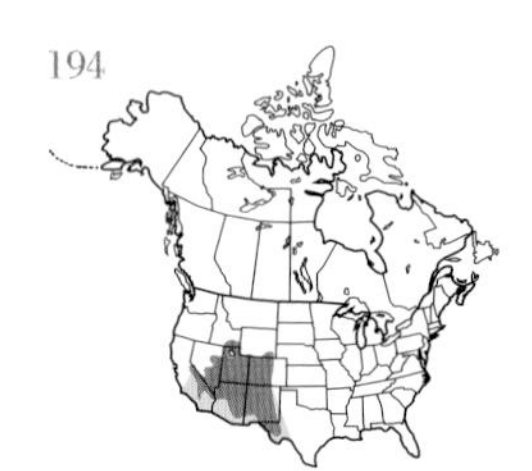

Virginia's Warbler

Oreothlypis virginiae

Although similar to a Nashville Warbler (p.193), a Virginia's Warbler has subtly different plumage and the two species' breeding ranges and habitats do not overlap. The sexes are dissimilar. Adult males have blue-gray upperparts, with a dark gray tail and wings, and a yellow rump. The head has a white eyering and a chestnut crown patch that is usually concealed. The pale underparts are whitish on the throat and belly, and yellow on the breast and undertail. Adult females are similar to an adult male but less colorful. In fall, all adults have plumage that is duller and browner than in spring. Immatures are similar to an adult female but lack a crown patch and have little or no yellow on the underparts. The legs are dark in all birds.

Virginia's Warbler is present as a breeding species to southwest U.S.A. mainly from April to August. It spends the rest of the year in Mexico. It bobs its tail up and down as it forages among foliage for insects.

female

male life-size

FACT FILE

LENGTH 4.75 in (12 cm)

FOOD Invertebrates

HABITAT Dry, scrubby oak or pine woodland, often on hillside slopes

STATUS Fairly common summer visitor

VOICE The two-part song comprises *swee-swee-swee-swee* and *swit-swit-swit-swit* phrases. Call is a sharp *chink*

Connecticut Warbler

Oporornis agilis

The Connecticut Warbler is a plump-bodied songbird. The sexes are dissimilar. Adult males have an olive-buff back, wings, and tail. The head has a gray hood, palest on the throat and darkest on the well-defined lower margin, and a white eyering. The underparts are otherwise bright yellow. Adult females are similar to adult males but less colorful overall and with a browner hood. Immatures are similar to an adult female but the hood and upperparts are buffish brown. The legs are pink in all birds.

The Connecticut Warbler is present as a breeding species in northern forests mainly from May to August. It spends the rest of the year in South America. Typically, it forages on the ground. Its secretive habits, combined with the relative inaccessibility of its favored habitats, mean it is hard to observe. Its presence is easiest to detect by listening for its song.

FACT FILE

LENGTH 5.75 in (14.5 cm)

FOOD Invertebrates

HABITAT Boggy northern forests

STATUS Widespread but scarce summer visitor

VOICE Song is a rich, chirpy *wee-chupa-chepa-chup-chup-wee*. Call is buzzing but seldom heard

MacGillivray's Warbler

Geothlypis tolmiei

MacGillivray's Warbler is the western, montane counterpart of Mourning Warbler (p.197). The sexes are dissimilar. Adult males have an olive-buff back, wings, and tail. There is a blue-gray hood, darkest on the lower margin and lores, and there is a broken white eyering. The underparts are otherwise bright yellow. Adult females are similar to an adult male but less colorful and with a pale gray hood. Immatures are similar to an adult female but with a mainly buffish-brown hood and a pale throat. The legs are pink in all birds.

MacGillivray's Warbler is present as a breeding species mainly from June to August. It spends the rest of the year in Central America. This secretive wood warbler favors dense thickets and undergrowth, making it a challenge to see.

FACT FILE

LENGTH 5.25 in (13.5 cm)

FOOD Invertebrates

HABITAT Dense mixed and conifer forests, usually near water

STATUS Common summer visitor

VOICE Song is in two parts, the first trilling, the second a series of *sweet-sweet-chooee* phrases. Call is a sharp *tzik*

Mourning Warbler

Geothlypis philadelphia

The Mourning Warbler is a plump ground-dwelling songbird. The sexes are dissimilar. Adult males have an olive-buff back, wings, and tail. There is a blue-gray hood, darkest on the lores and on the lower margin. The underparts are otherwise bright yellow. Adult females are similar to an adult male but the hood is uniformly pale gray. The lack of a pale eyering helps distinguish adults from an adult Connecticut Warbler (p.195). Immatures are similar to an adult female but with a drab gray head and neck, a yellowish throat, and an indistinct, broken pale eyering. The legs are pinkish in all birds.

The Mourning Warbler is present as a breeding species mainly from June to August. It spends the rest of the year in Central and South America. It is a secretive and unobtrusive bird, and a challenge to observe.

male

male life-size

FACT FILE

LENGTH 5.25 in (13.5 cm)

FOOD Invertebrates

HABITAT Dense scrub and secondary woodland thickets

STATUS Widespread and fairly common summer visitor

VOICE Song is usually in two parts, the first a set of buzzing whistles, the second a series of churring notes. Call is a thin *chit*

Kentucky Warbler

Geothlypis formosa

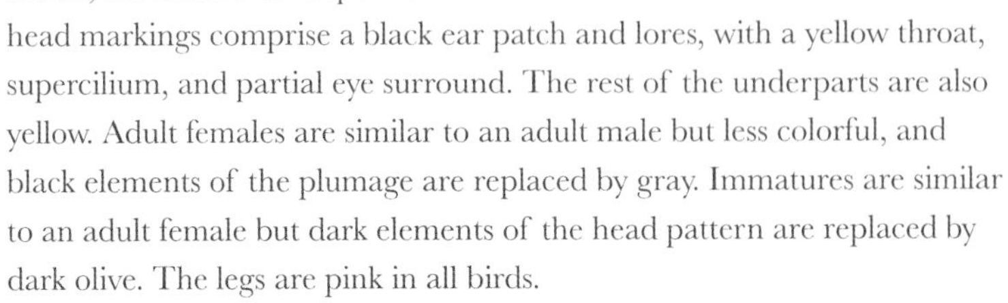
female

The Kentucky Warbler is a well-marked, rather short-tailed songbird. The sexes are subtly dissimilar. Adult males have olive-green upperparts extending to the rear of the crown; the forecrown is speckled black. The head markings comprise a black ear patch and lores, with a yellow throat, supercilium, and partial eye surround. The rest of the underparts are also yellow. Adult females are similar to an adult male but less colorful, and black elements of the plumage are replaced by gray. Immatures are similar to an adult female but dark elements of the head pattern are replaced by dark olive. The legs are pink in all birds.

The Kentucky Warbler is present as a breeding species in southeastern North America mainly from May to August. It spends the rest of the year in Central and South America. The species' secretive habits and the relative inaccessibility of its favored habitats means it is hard to observe.

male life-size

FACT FILE

LENGTH 5.25 in (13.5 cm)

FOOD Invertebrates

HABITAT Dense deciduous woodland, usually near water

STATUS Locally common summer visitor

VOICE Song is a series of churring *tseeup-tseeup* whistles. Call is a soft *tchup*

Common Yellowthroat

Geothlypis trichas

The Common Yellowthroat is a colorful little wood warbler. The sexes are dissimilar. Adult males have an olive-brown back and nape, and subtly darker wings and tail. The head has a broad black mask, defined above by a pale gray band, and below by the bright yellow throat; the yellow coloration extends to most of the underparts, although eastern birds typically have gray or buff flanks. Adult females are similar to their respective regional adult males but the black face mask is replaced by olive-gray. Immatures are similar to an adult female but the throat is less colorful. The legs are pink in all birds.

FACT FILE

LENGTH 5.5 in (14 cm)

FOOD Invertebrates

HABITAT Grassy habitats, often near water

STATUS Widespread and common summer visitor

VOICE Song is a warbled, whistling *weeter-weechertee-weechertee-wee*. Call is a sharp *tchet*

The Common Yellowthroat is present as a breeding species across much of North America, mainly from April to August. Although it can be found in southern U.S.A. in winter, most of the population moves to Central America outside the breeding season. It is a secretive species, and often its distinctive call is what alerts an observer to its presence, hidden among waterside grasses and sedges.

Hooded Warbler

Setophaga citrina

female

The Hooded Warbler is a distinctive songbird. The sexes are dissimilar. Adult males have an olive-green back with a subtly darker tail and wings. The yellow face is framed by black from the rear of the crown to the throat, and serves to highlight the dark eye. The underparts are otherwise yellow. Adult females are similar to an adult male but black elements of the head pattern are confined to the crown and nape. Immatures are similar to an adult female but dark elements of the head pattern are olive-green. The legs are pink in all birds.

The Hooded Warbler is present as a breeding species in eastern North America mainly from May to August. It spends the rest of the year in South America. As it forages for insects it often flicks its tail, revealing white on the outer feathers.

male
life-size

FACT FILE

LENGTH 5.25 in (13.5 cm)

FOOD Invertebrates

HABITAT Mature deciduous woodland and dense understory

STATUS Locally common summer visitor

VOICE Song is a series of whistled *weetu-weetu-weetu-wee-tee-tu* phrases. Call is a sharp *chip*

American Redstart

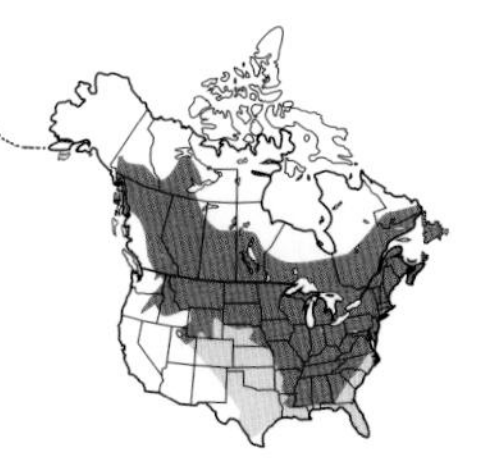

Setophaga ruticilla

female

The American Redstart is a well-marked and distinctive wood warbler. The sexes are dissimilar. Adult males have a black head, neck, chest, and upperparts, with orange patches on the wings and base of the tail. There is orange on the sides of breast, and the underparts are otherwise white. Adult females have gray-green upperparts and grayish-white underparts, with orange elements of the male's plumage replaced by yellow. Immatures are similar to an adult female, although the intensity and tone of the orange/yellow color varies. In their first year of life, birds in spring retain many immature plumage characters.

The American Redstart is present as a breeding species across much of central and eastern North America mainly from May to August. It spends the rest of the year in Central and South America. It feeds in an active manner, foraging for insects and fanning its tail as it goes, revealing the colorful patches at the base of the tail.

FACT FILE

LENGTH 5.25 in (13.5 cm)

FOOD Invertebrates

HABITAT Wide range of wooded habitats, including gardens

STATUS Widespread and common summer visitor

VOICE Song is variable but often a whistled *see-see-see-see-shweah*. Call is a shrill *chip*

male life-size

Kirtland's Warbler

Setophaga kirtlandii

Kirtland's Warbler is a well-marked songbird. The sexes are separable. Adult males have mostly blue-gray upperparts with dark streaks on the back and two white wingbars. The head pattern comprises a blue-gray hood, a broken white eyering, and a yellow throat. The rest of the underparts are also mostly yellow with dark streaks on the flanks; the color grades to white on the undertail. Adult females are similar to an adult male but less strikingly marked and less colorful. Immatures are similar to an adult female but the plumage colors are even duller, with buff fringes to the back feathers and wing coverts.

Kirtland's Warbler is present in a very restricted breeding range in Michigan, mainly from May to August; it is the rarest wood warbler in the region. The species spends the rest of the year in the Bahamas. It often forages at relatively low levels and pumps its tail up and down while feeding.

female

male life-size

FACT FILE

LENGTH 5.75 in (14.5 cm)

FOOD Invertebrates

HABITAT Jack Pine (*Pinus banksiana*) forests

STATUS Rare summer visitor

VOICE Song is a rich *tchew-tchew-tchwe-wee*. Call is a soft *tchip*

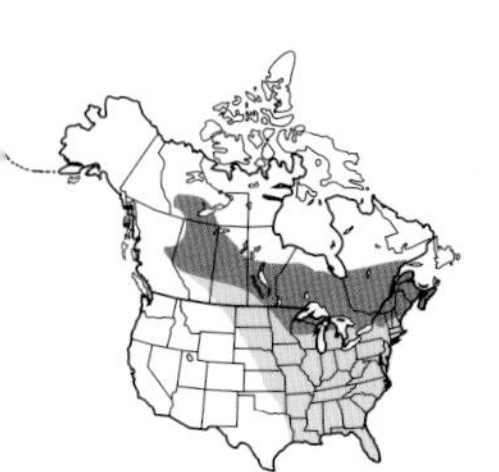

Cape May Warbler

Setophaga tigrina

The Cape May Warbler is an active little songbird. The sexes are dissimilar. Adult males have a streaked olive-green crown, nape, and back, with a pale rump and subtly darker tail and wings; there is a striking white wing patch. The yellow face is marked with chestnut ear coverts and a dark eye stripe. The underparts are yellowish with bold dark streaks on the breast and flanks; the undertail is white. Adult females are similar to an adult male but less colorful overall; the ear coverts are olive (not chestnut) and the wings have two white wingbars rather than a white panel. Adults of both sexes are duller in fall than in spring, and males usually have olive (not chestnut) ear coverts. Immatures are similar to their respective fall adults but the colors are even duller.

female

The Cape May Warbler is present as a breeding species across much of forested northern North America, mainly from May to August. It spends the rest of the year in the Caribbean region. The species usually feeds high in the treetops, making observation tricky. In plumages other than adult male, a pale yellow patch on the side of the neck provides a useful clue to identification.

male
life-size

FACT FILE

LENGTH 5 in (12.5 cm)

FOOD Invertebrates

HABITAT Northern forests

STATUS Widespread and common summer visitor

VOICE Song is a series of thin *tseet-tseet-tseet* notes. Call is a sharp *tzip*

Cerulean Warbler

Setophaga cerulea

female

The Cerulean Warbler is a beautiful but endangered songbird. The sexes are subtly dissimilar. Adult males have bright blue upperparts and mostly white underparts. There are dark streaks on the flanks and the wings show two white wingbars. Adult females are similar to an adult male but blue elements of the plumage are replaced by greenish blue. The subtly darker wings emphasize the white wingbars, and the underparts are suffused pale yellow. The head has a pale supercilium and the streaks on the flanks are indistinct. Immatures are similar to an adult female but with greenish-yellow upperparts.

The Cerulean Warbler is present as a rare breeding species in southeast North America, mainly from May to September. It spends the rest of the year in South America. Its population is declining due to habitat loss and degradation. It forages in the treetops, making it a challenge to observe.

male
life-size

FACT FILE

LENGTH 4.75 in (12 cm)

FOOD Invertebrates

HABITAT Mature deciduous forests with an open understory

STATUS Rare summer visitor

VOICE Song is a series of buzzing whistles, ending in a trill. Call is a soft *tsup*

Northern Parula

Setophaga americana

The Northern Parula is a colorful and well-marked wood warbler. The sexes are dissimilar. Adult males have mainly blue upperparts with a greenish patch on the back, and two white wingbars. The face has a broken white eyering. Below the yellow throat is a blue and orange breast band; the rest of the underparts grade from yellow to white on the undertail. Adult females are similar to an adult male but less colorful and without a breast band. Immatures are similar to an adult female but duller still.

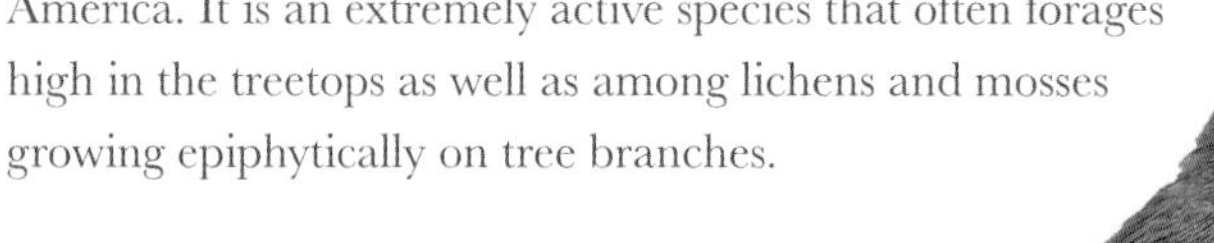

female

The Northern Parula is present as a breeding species in much of eastern North America mainly from April to August. It spends the rest of the year in Central America. It is an extremely active species that often forages high in the treetops as well as among lichens and mosses growing epiphytically on tree branches.

male
life-size

FACT FILE

LENGTH 4.5 in (11.5 cm)

FOOD Invertebrates

HABITAT Deciduous and mixed forest

STATUS Widespread and common summer visitor

VOICE Song is a squeaky trill that rises in tone. Call is a sharp *tzip*

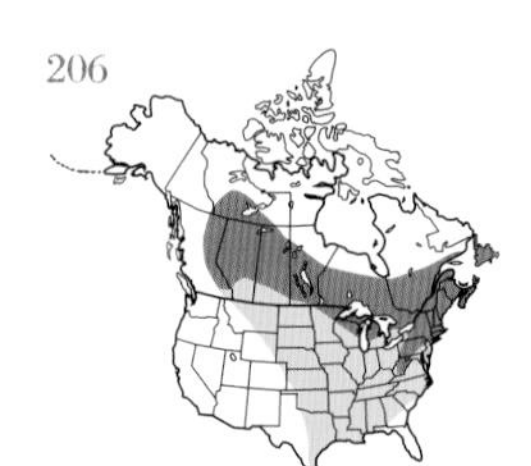

Magnolia Warbler

Setophaga magnolia

The Magnolia Warbler is a colorful species. In both sexes and in all plumages birds have a diagnostic tail underside, which is white with a broad, dark tip. Adult summer males have a mainly blackish back and wings with a broad white wing panel. The rump is yellow and the dark uppertail has a pale marginal band halfway along its length. A dark mask links to the dark back; note the white supercilium and blue-gray crown. The underparts are mostly bright yellow with a black chest band and streaked flanks; the undertail coverts are white. Adult summer females are similar to a summer male but black elements of the body plumage are gray, not black. In fall, adult plumage in both sexes is similar to that of an adult summer female but without the white supercilium, and with reduced white on the wings. Immatures are similar to non-breeding adults but the underparts are not streaked.

The Magnolia Warbler is present as a breeding species in North America from late May to August. It spends the rest of the year in Central America. While nesting, the species is associated with mature and sometimes large trees; however, it often feeds relatively low down in the shrub understory, allowing good views to be obtained.

FACT FILE

LENGTH 5 in (12.5 cm)

FOOD Invertebrates

HABITAT Northern conifer forests

STATUS Widespread and locally common summer visitor

VOICE Song is a whistled *zwee-zwee-zwee-zweep*. Call is a thin *tsic*

Bay-breasted Warbler

Setophaga castanea

female

The Bay-breasted Warbler is a well-marked songbird, males being particularly colorful. The sexes are dissimilar. Adult summer males have a dark, streaked back and nape, and dark wings with two white wingbars. The crown, throat, chest, and flanks are chestnut, the face is black, and the side of the neck is buff. The underparts are otherwise creamy white. Adult summer females recall an adult male, but chestnut and black elements of the plumage are mainly replaced with gray-buff. In fall, adults of both sexes are similar to an adult summer female but less heavily streaked. Immatures are similar to a fall adult female, but duller still, and with only faint streaking on the back, and buff underparts that include the undertail. The legs are dark in all birds. Immature Blackpoll Warblers (p.211), which are similar to immature Bay-breasted Warblers, have pinkish-orange legs.

The Bay-breasted Warbler breeds across much of forested northern North America mainly from May to August. It spends the rest of the year in Central America. Its breeding success is linked to the abundance or otherwise of spruce budworm.

male life-size

FACT FILE

LENGTH 5.5 in (14 cm)

FOOD Invertebrates, notably spruce budworms (larvae of moths in the genus *Choristoneura*)

HABITAT Mature spruce forests

STATUS Widespread and common summer visitor

VOICE Song is a series of five or more high-pitched whistling *tswee* notes. Call is a thin *tssip*

Blackburnian Warbler

Setophaga fusca

female

The Blackburnian Warbler is a strikingly marked songbird. The sexes are dissimilar. Adult summer males have a black back, nape, and crown. The wings are black with a broad white patch on the coverts and pale edges to the flight feathers. The mainly black tail has white patches on the outer feathers. The face is orange with a black patch through the eye, and the throat is orange (unique among wood warblers). The breast is yellowish with dark streaks on the flanks, and the underparts are otherwise white with black streaks. Adult females are similar to an adult male but black elements of the plumage are streaked grayish yellow and the face is yellow, not orange. Adults in fall are similar to a summer female but less colorful, and immatures are similar but duller and paler still.

The Blackburnian Warbler is present as a breeding species in forests in northern North America and eastern mountain ranges mainly from May to August. It spends the rest of the year in South America. Typically it forages in the treetops, making observation a bit of a challenge.

male life-size

FACT FILE

LENGTH 5 in (12.5 cm)

FOOD Invertebrates

HABITAT Northern and montane conifer forests

STATUS Widespread and common summer visitor

VOICE Song is often a series of very high-pitched notes that ends in a trill. Call is a sharp *tsik*

Yellow Warbler

Setophaga petechia

female

The Yellow Warbler is an aptly named songbird. The sexes are subtly dissimilar in any given season, but adults are subtly brighter in spring than fall. Adult males are bright yellow, the subtly darker wings showing two pale wingbars. The breast and flanks have reddish streaks. Regional variation exists (represented by several subspecies), with northern birds being darkest and most intensely marked, and southwestern birds being plainest overall. Adult females recall their respective regional male but are more uniformly yellow overall, and almost unstreaked below. Immatures are similar to an adult female but much less colorful; some are gray-buff.

The Yellow Warbler is present as a breeding species across most of North America, mainly from April to August. It spends the rest of the year in Central and South America. Typically it forages for insects in low vegetation, making it easy to observe.

western male

northern male
life-size

FACT FILE

LENGTH 5 in (12.5 cm)

FOOD Invertebrates

HABITAT Scrubby willow thickets and secondary-growth woodland

STATUS Widespread and common summer visitor

VOICE Song is a whistling *swee-swee-swee-swee-swit-su-swee*. Call is sharp *tchip*

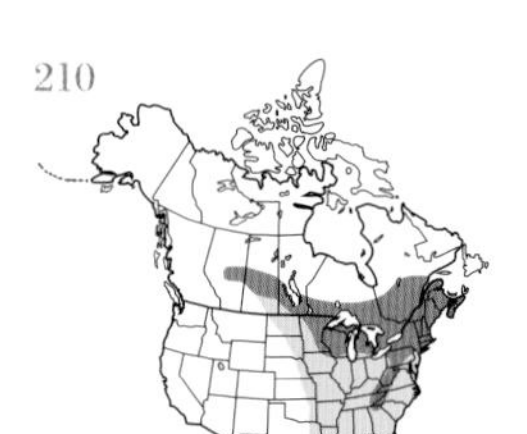

Chestnut-sided Warbler

Setophaga pensylvanica

female

The Chestnut-sided Warbler is a distinctively marked songbird. The sexes are subtly dissimilar. Adult summer males have a streaked black and white back and nape, and black wings with two white wingbars. The head pattern comprises a yellow crown, a black eye stripe and "mustache," and an otherwise white face. The underparts are mostly white but with chestnut flanks. Adult summer females are similar but less colorful overall, with a streaked crown and less chestnut on the flanks. Adults in fall are less colorful than their summer counterparts, and have much less chestnut on the flanks, gray-green upperparts, a gray face and throat, and a white eyering. Immatures are similar to a fall adult but with no chestnut on the flanks.

The Chestnut-sided Warbler is present as a breeding species in northeast North America, mainly from April to August. It spends the rest of the year in Central America. It is an active species that often forages for insects low in bushes, making it easy to observe.

immature

male life-size

FACT FILE

LENGTH 5 in (12.5 cm)

FOOD Invertebrates

HABITAT Secondary-growth woodland

STATUS Widespread and common summer visitor

VOICE Song is a descending series of sweet whistles, *swee-si-si, tsi-tsuu swee-sa*. Call is a sharp *tchhup*

Blackpoll Warbler

Setophaga striata

female

The Blackpoll Warbler is a strikingly marked little songbird. The sexes are dissimilar. Adult summer males have a dark, streaked back and nape, with a black cap, white face, and black malar stripe. The black wings have two bold white wingbars. The underparts are white with bold black streaks on the flanks. Adult summer females appear grubby by comparison, and the head is streaked olive-gray except for the pale throat and dark malar stripe. Adults in fall are similar to a summer female. Immatures are similar to a fall adult, but the upperparts are olive-yellow and the face is brighter yellow. The underparts are olive-yellow, subtly streaked on the flanks and grading to white on the undertail. All birds have orange legs and feet. Compared to an immature Bay-breasted Warbler (p.207), an immature Blackpoll Warbler has orange-yellow (not dark) legs and feet, and a white (not buff) undertail.

The Blackpoll Warbler is present as a breeding species in northern forests of North America mainly from June to August. It spends the rest of the year in northern South America. Birds feed in a relatively slow, deliberate manner.

male
life-size

FACT FILE

LENGTH 5.5 in (14 cm)

FOOD Invertebrates

HABITAT Boreal forests

STATUS Widespread and common summer visitor

VOICE Song is a series of very high-pitched *tsi-tsi-tsi…* notes. Call is a sharp *chip*

Black-throated Blue Warbler

Setophaga caerulescens

The Black-throated Blue Warbler has strikingly different male and female plumages. Adult summer males have mainly dark blue upperparts with blackish wings that show a white patch at the base of the primaries. The blackish tail has white patches on the outer feathers. The face, throat, and flanks are black with a distinct division from the otherwise white underparts. Adult females have dark buffish-brown upperparts, the subtly darker wings showing a white patch at the base of the primaries. The underparts are pale buffish yellow, grading to whitish on the undertail. The head has a pale supercilium and crescent below the eye. Immatures are similar to an adult female, but the pale wing patch is indistinct or absent and the supercilium and eye crescent are less distinct.

The Black-throated Blue Warbler is present as a breeding species in north-central North America mainly from May to August. It spends the rest of the year in the Caribbean region. It often forages for insects in relatively low shrubs, when it is easy to observe.

FACT FILE

LENGTH 5.25 in (13.5 cm)

FOOD Mainly invertebrates, but also berries in fall

HABITAT Upland deciduous and mixed forests

STATUS Locally common summer visitor

VOICE Song comprises a series of squeaky *zhee-zerr-zhree*... notes. Call is a sharp *tuuk*

male life-size

female

Palm Warbler

Setophaga palmarum

western male

Palm Warbler sexes are very similar to one another, although males are usually subtly brighter than females. To complicate matters, birds from the east of the breeding range are brighter than western birds but all birds occupy the same range in winter. Eastern summer adults have faintly streaked olive-brown upperparts and darker wings with two pale wingbars. The head has a chestnut crown, dark eye stripe, yellow supercilium, olive cheeks, and a yellow throat. The underparts are otherwise yellow with rufous streaks on the flanks. Western summer adults have a grayer back; yellow on the underparts is confined to the throat and undertail. Adults in fall, and immatures, are less colorful than their respective spring adult counterparts, and lack the rufous crown and streaks on the flanks. Eastern birds have a yellowish supercilium and yellow-suffused underparts, while western birds have a white supercilium and underparts that are gray except for the yellow undertail.

The Palm Warbler breeds in northern North America mainly from May to August. It spends the rest of the year in southeast U.S.A. and the Caribbean region. It often feeds on the ground and wags its tail as it moves.

eastern male life-size

FACT FILE

LENGTH 5.5 in (14 cm)

FOOD Invertebrates

HABITAT Northern conifer forests in summer; wide range of woodland in winter

STATUS Widespread and common summer visitor; fairly common in winter in the southeast

VOICE Song is a trilling series of buzzing notes. Call is a sharp *tchik*

Pine Warbler

Setophaga pinus

The Pine Warbler is aptly named because it is invariably associated with pine trees. The sexes are subtly dissimilar. Adult males have mostly olive-yellow upperparts, but darker wings that have two white wingbars and white feather margins. The head has indistinct markings comprising a faint supercilium in front of the eye and an incomplete yellow eyering. The underparts are mainly yellowish, grading to white on the belly and undertail; there are faint dark streaks on the flanks. Adult females are similar but less colorful and less well marked. Immatures are duller and less colorful than their respective adults.

The Pine Warbler is present as a breeding species in northern forests mainly from April to September. In southeast U.S.A. it is present year-round, numbers there being boosted in winter by migrants from the north. It sometimes forages in low scrub, and even on the ground, when observation of the species is easy.

FACT FILE

LENGTH 5.25 in (13.5 cm)

FOOD Invertebrates

HABITAT Pine forests

STATUS Widespread and common summer visitor, present year-round in the southeast of its range

VOICE Song is a series of tuneful trilling notes. Call is buzzing *tzeep*

Yellow-throated Warbler

Setophaga dominica

eastern male

The Yellow-throated Warbler is a well-marked little songbird. The sexes are subtly dissimilar. Adult males have a blue-gray back, nape, and rear crown, darkening to black on the forecrown. The dark wings have two white wingbars, and the tail is dark. The head pattern comprises a black face with a pale supercilium (pure white in western birds, tinged yellow in front of the eye in eastern birds), a white lower "eyelid," and a white patch behind the ear coverts. The throat and chest are yellow, and the underparts are otherwise white with dark streaks on the flanks. Adult females are similar to their respective adult males, but with a grayer forecrown and fainter stripes on the flanks. Immatures are similar to an adult female, but black elements of the plumage are paler still and the underparts are suffused buff.

The Yellow-throated Warbler is present as a breeding species in eastern U.S.A. mainly from April to August. At other times of the year it is found mainly in Central America, although small numbers spend the winter in southeast U.S.A. It forages in a relatively slow and deliberate manner, looking for insects among leaves.

FACT FILE

LENGTH 5.25 in (13.5 cm)

FOOD Invertebrates

HABITAT Range of wooded habitats, from inundated forests to dry coniferous woodland

STATUS Widespread and common summer visitor

VOICE Song is series of thin, whistled *tsi-tsi-tsi-tsu-tsu* notes. Call is a thin *tsip*

western male life-size

Yellow-rumped Warbler

Setophaga coronata

Myrtle Warbler

Audubon's Warbler

All Yellow-rumped Warblers have a yellow rump (a feature also seen in Magnolia (p.206) and Cape May (p.203) warblers) and a yellow flank patch. The sexes are dissimilar and the male's breeding plumage varies significantly across its range. Northern and eastern populations are called "Myrtle Warblers," while western populations are referred to as "Audubon's Warblers." A summer adult male Myrtle has mostly blue-gray upperparts, streaked on the back, with two white wingbars and a yellow crown stripe. The head is dark but with a thin white supercilium and white throat. The otherwise white underparts have a dark breast band and dark streaks on the flanks. A male Audubon's is similar, but the throat is yellow and the white color on the wings is more extensive. A summer adult female Myrtle has brown upperparts, an indistinct yellow crown patch, two faint wingbars, a white throat, and otherwise streaked white underparts. A female Audubon's is similar, but grayer overall and with a yellow-flushed throat. Fall adults are duller than their summer counterparts. Immatures are like their respective fall females but warmer buff overall and without a crown patch; the throat is buff in Audubon's but white in Myrtle, and the latter also has a thin, pale supercilium.

Audubon's Warbler, female

FACT FILE

LENGTH 5.25 in (13.5 cm)

FOOD Invertebrates and berries

HABITAT Boreal forests, and montane conifer and mixed forests farther south

STATUS Widespread and common summer visitor; widespread in winter in the south

VOICE Song is a series of whistled trills. Call is a soft *chep* or *chik*

The Yellow-rumped Warbler is present as a breeding species across northern and northwestern North America mainly from May to August. Birds migrate south in fall, with the winter range extending from southern U.S.A. to Central America.

Audubon's Warbler, male life-size

Myrtle Warbler, female

Myrtle Warbler, male life-size

Prairie Warbler

Setophaga discolor

Despite its name, the Prairie Warbler is not a grassland species. The sexes are subtly dissimilar. Adult males have olive-yellow upperparts, and subtly darker wings with two pale yellowish wingbars. The head pattern comprises a yellow face adorned with a dark eye stripe linked to a dark semicircle below the eye. The throat and underparts are mostly bright yellow with black streaks on the flanks and sides of the neck. Adult females are similar to an adult male but the face markings are indistinct. Immatures are duller and paler overall than their respective adults.

male life-size

The Prairie Warbler is present as a breeding species in eastern North America mainly from May to August. At other times of the year it is found in the Caribbean region, with some birds spending the winter in Florida. While foraging for insects, the tail is usually pumped up and down.

FACT FILE

LENGTH 4.75 in (12 cm)

FOOD Invertebrates

HABITAT Scrub-colonized grassland and young-growth woodland

STATUS Widespread and locally common summer visitor; local in winter

VOICE Song is a series of buzzing *tsu, tsu, tsi-si-si-si* notes that rise in pitch. Call is a sharp *tchip*

female

Grace's Warbler

Setophaga graciae

Grace's Warbler is a southwestern specialty. The sexes are subtly dissimilar. Adult males have blue-gray upperparts with dark streaks on the back and crown, and dark wings with two white wingbars. The face pattern comprises dark ear coverts and lores, and a yellow supercilium and spot below the eye. The throat and breast are yellow, grading to white on the rest of the underparts, with dark streaks on the flanks. Adult females are similar to an adult male, but less colorful and paler overall, and with unstreaked upperparts. Immatures are similar to an adult female but paler still, with browner upperparts and a yellow suffusion on the flanks.

Grace's Warbler is present as a breeding species, mainly in Arizona and New Mexico, from May to August, where it favors stands of Ponderosa Pines (*Pinus ponderosa*). It spends the rest of the year in Central America. It often feeds high in the tree canopy, making observation a bit of a challenge.

male

FACT FILE

LENGTH 5 in (12.5 cm)

FOOD Invertebrates

HABITAT Conifer forests

STATUS Local summer visitor

VOICE Song is an accelerating series of chattering notes. Call is a soft *chip*

male
life-size

Black-throated Gray Warbler

Setophaga nigrescens

The Black-throated Gray Warbler is a well-marked little songbird. The sexes are subtly dissimilar. Adult males have a streaked, dark gray back and wings with two white wingbars. The mainly black head is patterned with a white supercilium and malar stripe, and a tiny yellow spot in front of the eye. The throat and breast are also black, and the underparts are otherwise white with dark-streaked flanks. The tail is dark above with white outer feathers, and mostly white below. Adult females are similar to an adult male, but the throat is white with faint dark streaking; black elements of the head pattern are grayer, and the back is paler. Immatures are similar to an adult female but grayer still, with a less distinct breast band and a buff suffusion to the underparts.

The Black-throated Gray Warbler is present as a breeding species in western North America mainly from May to August. It spends the rest of the year in Mexico.

female

male life-size

FACT FILE

LENGTH 5 in (12.5 cm)

FOOD Invertebrates

HABITAT Open conifer and mixed woodland

STATUS Widespread and common summer visitor

VOICE Song is a rapid series of buzzing notes, such as *whzz-tzee, whzz-tzee, whzz-tzee, whzz-zoo*. Call is a thin *tsip*

Townsend's Warbler

Setophaga townsendi

female

Townsend's Warbler is a stunning little songbird. The sexes are subtly dissimilar. Adult males have a streaked olive-yellow back and rump, blackish wings with two white wingbars, and a blackish tail with white outer feathers. The head has a dark crown, a dark "mask" with a yellow patch below the eye, and a black throat. The chest is black and the underparts are otherwise yellow on the breast and flanks (which have dark streaks), grading to white on the belly and undertail. Adult females are similar to an adult male but paler overall, with black elements of the head markings replaced by olive-gray. Immatures are similar to an adult female but less colorful, and with much fainter markings on the breast.

Townsend's Warbler is present as a breeding species in Pacific Northwest conifer forests mainly from May to August. It spends the rest of the year in coastal California and Mexico, where pine forests are also favored. It usually forages for insects high in the treetops.

male life-size

FACT FILE

LENGTH 5 in (12.5 cm)

FOOD Invertebrates

HABITAT Conifer forests

STATUS Locally common summer visitor

VOICE Song is a buzzing *weeze-weeze-weeze-weZhe*. Call is a sharp *tsic*

Hermit Warbler

Setophaga occidentalis

female

The Hermit Warbler is a distinctive songbird. The sexes are subtly dissimilar. Adult males have dark gray upperparts with subtly darker wings showing two white wingbars. The tail is dark above with white outer feathers, and white below. The head pattern comprises a yellow face, surrounded by a dark-speckled nape and black throat and chest. The underparts are white with very faint dark streaking. Adult females are similar to an adult male but the upperparts are paler, and the ear coverts are tinged gray. Immatures are similar to an adult female but the upperparts are dull olive-brown.

The Hermit Warbler is a Pacific Northwest specialty and is present in montane conifer forests as a breeding species mainly from May to August. It spends the rest of the year in Central America. It is a notoriously secretive species that usually feeds high in the treetops.

male
life-size

FACT FILE

LENGTH 5.25 in (13.5 cm)

FOOD Invertebrates

HABITAT Upland conifer forests

STATUS Locally common summer visitor

VOICE Song is a rattling *wezee-wezee-wezee*... Call is a thin *tzip*

Golden-cheeked Warbler

Setophaga chrysoparia

The Golden-cheeked Warbler is an aptly named species. The sexes are dissimilar. Adult males have a black back, nape, crown, and throat; this color contrasts with, and offsets, the yellow face, which has a dark eye stripe. The black wings have two white wingbars, and the tail is dark above with white outer feathers, and white below. The underparts are otherwise white with black streaks on the flanks. Adult females are similar to an adult male but the throat is yellowish and flecked with black, and other black elements of the male's plumage are dark gray. Immatures are similar to an adult female but lack the dark chest band and have only faint streaks on the flanks.

male life-size

The Golden-cheeked Warbler is present as a breeding species in its restricted range in central Texas mainly from April to June. It spends the rest of the year in Central America.

FACT FILE

LENGTH 5.25 in (13.5 cm)

FOOD Invertebrates

HABITAT Mixed oak and juniper woodland

STATUS Rare summer visitor

VOICE Song is a buzzing *drr-zee, drr-zee, drr-zee*… Call is a soft *tsip*

female

Black-throated Green Warbler

Setophaga virens

The Black-throated Green Warbler is a distinctive black-faced warbler. The sexes are dissimilar. Adult males have an olive-yellow back, nape, and crown. The wings are dark with two white wingbars. The tail is dark above with white outer feathers, and white below. The face is yellow with olive-buff ear coverts. The throat and chest are black, and the underparts are otherwise mainly white with black streaks on the flanks, and a subtle yellow suffusion on the chest and flanks. Adult females are similar to an adult male but the throat is mottled yellow. Immatures are similar to an adult female but black feathering on the underparts is replaced by olive-brown.

The Black-throated Green Warbler is present as a breeding species in northern North America mainly from May to August. It spends the rest of the year in the Caribbean region. The species is an active feeder and sometimes hovers while searching for insects.

female

male life-size

FACT FILE

LENGTH 5 in (12.5 cm)

FOOD Invertebrates

HABITAT Conifer and mixed forests

STATUS Widespread and common summer visitor

VOICE Song is a series of buzzing whistles such as *tzee-zee-zee-tzur-zee*. Call is a soft *tsip*

Canada Warbler

Cardellina canadensis

female

The Canada Warbler is a colorful little songbird. The sexes are dissimilar. Adult males have dark blue-gray upperparts that are subtly darkest on the wings and forecrown. There is a clear division between the dark cap and the yellow throat, and the eye is highlighted by a white eyering, which continues as a yellow line on the lores. The yellow underparts have dark streaks on the breast. Adult females resemble an adult male, but the upperparts are paler gray and the dark streaking on the breast is much fainter. Immatures are similar to an adult female but with a paler forehead and even less distinct streaking on the breast.

The Canada Warbler is present as a breeding species in northern forests mainly from June to early August; at this time it is often found near water. It spends the rest of the year in northern South America.

FACT FILE

LENGTH 5.25 in (13.5 cm)

FOOD Invertebrates

HABITAT Damp conifer and mixed forests with dense undergrowth

STATUS Widespread and fairly common summer visitor

VOICE Song is a shrill *tchip, chu-ptti, chu-ptti, chu-ptti*. Call is a sharp *tchiup*

male life-size

Wilson's Warbler

Cardellina pusilla

FACT FILE

LENGTH 4.75 in (12 cm)

FOOD Invertebrates

HABITAT Damp forests with dense undergrowth

STATUS Widespread and common summer visitor

VOICE Song is a whistled *wee-chee-chee-chee-chee*... Call is a sharp *tchip*

Wilson's Warbler is a small, plump songbird. The sexes are dissimilar. Adult males have olive-yellow upperparts that are darkest on the wings. The yellow face contrasts with the black eye and center of the crown. The underparts, including the throat, are yellow. Adult females are similar to an adult male but the crown is dark olive, not black. Immatures are similar to an adult female but the crown is paler olive-brown.

female

Wilson's Warbler is present as a breeding species in north and west North America, mainly from May to August. It spends the rest of the year in Central America. The species is a very active feeder and sometimes makes flycatching forays from a perch.

Red-faced Warbler

Cardellina rubrifrons

male

The Red-faced Warbler is an unmistakable little songbird. The sexes are subtly dissimilar. Adult males have a gray back with a white rump and white patch on the nape. The wings are gray with a white wingbar and the tail is dark gray. The head is adorned with a red face and a black hood. The underparts are whitish, washed pale gray on the breast. Adult females are similar to an adult male, but the color on the face is duller and orange-toned. Immatures are similar to an adult female but the back is buffish gray.

The Red-faced Warbler is a southwestern specialty, present in its range in Arizona and New Mexico mainly from May to August. It spends the rest of the year in Central America. Although it forages actively for insects, it will also undertake flycatching forays from a perch.

FACT FILE

LENGTH 5.5 in (14 cm)

FOOD Invertebrates

HABITAT Montane oak and pine forests

STATUS Locally common summer visitor

VOICE Song is a whistled *tsee-wee, tsee-we, tsee-wee, tsiweo.* Call is a sharp *tchup*

male
life-size

Painted Redstart

Myioborus pictus

The Painted Redstart is a stunning wood warbler. The sexes are dissimilar. Adult males are mostly black except for the white crescent below the eye, the broad white patch on the wings, the white outer-tail feathers, and the bright red belly. Adult females are similar to an adult male but the belly is dark gray, not red. Immatures are similar to an adult female until fall, when adult plumage differences become apparent in the case of males.

The Painted Redstart is another southwest specialty, present in Arizona and New Mexico mainly from March to September. It spends the rest of the year in Central America. The white markings on the wings and tail are seen to good effect when perched birds fan their tails and flick their wings.

male

male life-size

FACT FILE

LENGTH 5.75 in (14.5 cm)

FOOD Invertebrates

HABITAT Upland pine and oak woodland

STATUS Locally common summer visitor

VOICE Song is a rapid *weepa-weepa-wee-pee*. Call is a thin *schree*

Yellow-breasted Chat

Icteria virens

female

The Yellow-breasted Chat is a thick-billed, plump-bodied songbird. The sexes are subtly dissimilar. Adult males have a gray-brown back, wings, and tail, with eastern birds being grayer overall than their western counterparts. The head is patterned with a gray hood, a partial white eyering, and a white supercilium in front of the eye. A white stripe separates the hood from the yellow throat and this color continues on the underparts to the undertail, which is white. Adult females are similar to an adult male but with less striking head markings. Immatures are similar to an adult female but duller overall.

The Yellow-breasted Chat is present as a breeding species across the southern half of North America mainly from May to August. It spends the rest of the year in Central America. It is very secretive and hence easily overlooked; observation is made trickier by the dense nature of its favored habitat.

male life-size

FACT FILE

LENGTH 7.5 in (19 cm)

FOOD Invertebrates

HABITAT Dense colonizing scrub, usually on damp ground

STATUS Widespread but local summer visitor

VOICE Song is varied, including repeated chattering phrases, fluty notes, and whistles. Call is a harsh *tchup*

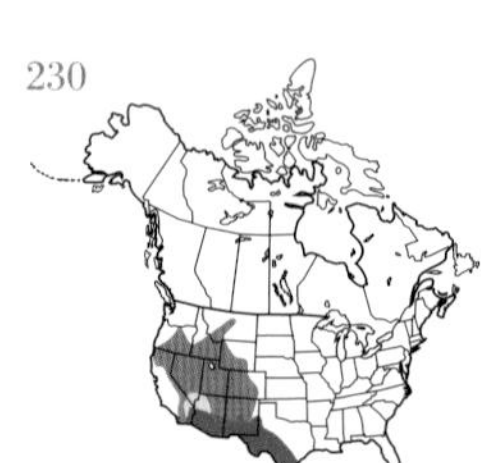

Green-tailed Towhee

Pipilo chlorurus

The Green-tailed Towhee is a long-tailed ground-dwelling songbird. The sexes are similar. Adults have an olive-green back. The wings and tail are also olive-green overall but yellow feather edges give them a much brighter appearance. The head is patterned with a chestnut crown, a white patch in front of the eye, and a white throat and malar stripe separated by a dark line. The face, neck, and underparts are mostly gray, grading to pale buff on the belly and undertail. Juveniles are brown and streaked, darker above than below, and with throat markings that recall those of an adult.

FACT FILE

LENGTH 7.25 in (18.5 cm)

FOOD Invertebrates and fallen seeds

HABITAT Dense chaparral scrub, on upland slopes in summer but at lower altitudes in winter

STATUS Locally common summer visitor; local in winter

VOICE Song is a series of chattering and whistling phrases, which starts with sharp *tchup* notes and ends in a trill. Call is a catlike *meow*

adult life-size

The Green-tailed Towhee is present as a breeding species in southwest North America, mainly from April to September. It spends the rest of the year mainly in Mexico, although the species' winter range also extends to the border regions of southern U.S.A. It forages on the ground, usually in dense cover, making it hard to observe. However, in spring males sometimes sing from exposed perches.

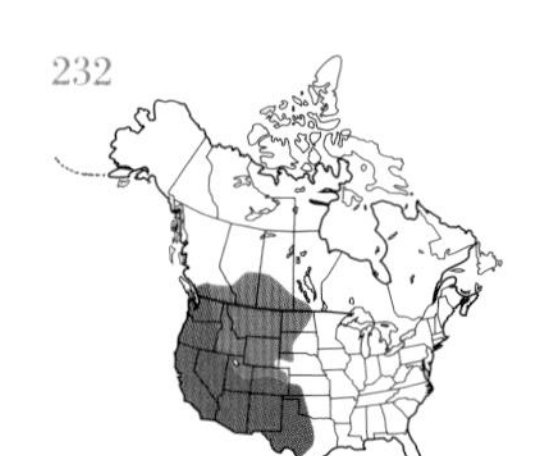

Spotted Towhee

Pipilo maculatus

The Spotted Towhee is well-marked long-tailed songbird. The sexes are subtly dissimilar and regional plumage variation also exists. Adult males have a black hood, upperparts, and tail; the extent of white markings on the wings and back varies according to the subspecies. Interior birds have white spots on the back, two white wingbars, and other white feather edges. At the other extreme, Pacific Northwest birds have almost unspotted black backs and reduced white wing markings. All birds have reddish-orange flanks, a white center to the belly, a buff undertail, and a red eye. Adult females are similar to their respective regional males but black elements of the plumage are dark brown. Juveniles are brown and streaked with two pale wingbars.

Pacific Northwest male

interior female

The Spotted Towhee has a complicated North American distribution. It is present year-round across much of the west of its range, but birds that breed in the north and interior are present there mainly from April to September, migrating south in fall. The species' winter range extends to Mexico. It often feeds in dense cover and is not always easy to see well.

interior male
life-size

FACT FILE

LENGTH 7.5 in (19 cm)

FOOD Invertebrates and fallen seeds

HABITAT Chaparral and scrubby woodland, and dense brush

STATUS Widespread and common resident and partial migrant

VOICE Song is a series of trill phrases, sometimes preceded by whistling notes. Call is a rasping mew

Eastern Towhee

Pipilo erythrophthalmus

The Eastern Towhee is a long-tailed bird that scratches for food among leaf litter. The sexes are dissimilar. Adult males have a black hood, upperparts, and tail, the wings with white margins to the flight feathers and white at the base of the primaries. The flanks are reddish orange, the breast and center of the belly are white, and the undertail is buff. Most birds have red eyes but in Florida birds they are yellow. Adult females are similar to an adult male but black elements of the plumage are dark brown. Juveniles are brown and streaked with two pale wingbars.

male life-size

FACT FILE

LENGTH 7.5 in (19 cm)

FOOD Invertebrates and fallen seeds

HABITAT Brush and dense scrub

STATUS Locally common summer visitor, present year-round in the south of its range

VOICE Song is a whistling *sweet-too-tee*, followed by a trill. Call is a shrill *tchewik*

The Eastern Towhee occurs year-round in southeast North America, but farther north it is a summer visitor, present there mainly from May to September. Like other towhee species, it feeds by scratching the ground using both feet together, and flies low to the ground, with long glides and short bouts of rapid wing-flapping.

female

Florida male

Abert's Towhee

Melozone aberti

Abert's Towhee is a long-tailed songbird that is rather plain-looking except for its dark face. The sexes are similar. Adults have sandy-brown upperparts that are darkest on the wings and tail. The head and underparts are mostly pinkish buff except for the black face, which contrasts with the pale bill; the undertail is pinkish orange. Juveniles are similar to an adult but have faint streaking on the underparts and indistinct pale wingbars.

Abert's Towhee is present year-round in its favored arid habitats in southwest U.S.A. Typically it keeps to dense cover, scratching the ground for insects and seeds; birds can sometimes be located by listening for feeding activity.

FACT FILE

LENGTH 9.5 in (24 cm)

FOOD Invertebrates and seeds

HABITAT Desert woodlands, particularly riverside locations

STATUS Locally common resident

VOICE Song is series of squeaky *pik* notes that ends in a trill. Call is shrill *peek*

adult life-size

adult

Canyon Towhee

Melozone fusca

The Canyon Towhee is a rather pale ground-dwelling songbird. The sexes are similar. Adults have gray-buff body plumage, darker above than below, and a subtly darker tail. The head has a reddish-brown cap and a pale gray-buff face with a yellowish eyering. The buffish-yellow throat is framed by a dark lateral line and a row of dark streaks on the breast, below which is a dark central spot. The rest of the underparts are gray-buff, palest on the belly and with a pinkish-orange undertail. Juveniles are similar to an adult but with a less colorful crown, faint streaking on the underparts, and indistinct pale wingbars.

The Canyon Towhee is present year-round in southwest U.S.A. It feeds on the ground, often in cover, and consequently is easy to overlook.

adult

adult life-size

FACT FILE

LENGTH 8.5 in (21.5 cm)

FOOD Invertebrates and seeds

HABITAT Arid canyons

STATUS Locally common resident

VOICE Song is a series of fluty whistles, with some repetition. Call is a shrill *chee-lup*

California Towhee

Melozone crissalis

The California Towhee is a rather plain-looking long-tailed songbird with a pointed bill. The sexes are similar. Adults have gray-brown upperparts. The head has a yellow-buff throat and lores; the former is faintly streaked and defined below by a row of dark streak-like spots on the upper breast. The underparts are otherwise gray-buff, palest on the belly and with a pinkish-orange undertail. Juveniles are similar to an adult but with additional faint streaking on the breast, and faint pale wingbars.

The California Towhee is present year-round along the coastal belt of California, its range extending south through Baja California in Mexico. Where suitable habitats exist, it has adapted to suburban locations, including parks and gardens.

adult

adult life-size

FACT FILE

LENGTH 9 in (23 cm)

FOOD Invertebrates and seeds

HABITAT Chaparral woodland with dense undergrowth, and scrub

STATUS Common resident

VOICE Song is a series of squeaky notes, sometimes delivered as a duet by a pair. Call is a thin *chink* note

Rufous-crowned Sparrow

Aimophila ruficeps

The Rufous-crowned Sparrow has rather nondescript plumage except for its head markings. The sexes are similar but regional variation exists. Adults have a streaked gray-brown back, a subtly darker tail, and dark wings with rufous margins to the inner flight feathers. The head has a rufous crown with a pale central stripe, a gray face with a white eyering, a brown stripe behind the eye, and a pale supercilium in front of the eye. There is a dark malar stripe and a "mustache" that is buff in coastal birds but white in interior birds. All birds have pale gray underparts. Juveniles are similar to adults but more streaked.

The Rufous-crowned Sparrow is present year-round in southern and southwest U.S.A. It is an unobtrusive species but fortunately it often scans its surroundings from a prominent boulder; males sing from similar spots in spring.

coastal adult

interior adult life-size

FACT FILE

LENGTH 6 in (15 cm)

FOOD Invertebrates and seeds

HABITAT Dry rocky slopes with sparse grasses

STATUS Locally common resident

VOICE Song is a rattling *chip-chip-chip chip-chit*… Call is a nasal *dneer*

Cassin's Sparrow

Peucaea cassinii

Cassin's Sparrow is a furtive songbird with plumage that lacks any really distinctive features. The sexes are similar. Adults have a streaked gray-brown back, the individual feathers with rufous centers and pale margins. The streaked brown crown has a pale central stripe and the pale gray face includes a pale gray supercilium. The whitish throat is framed by a thin dark line, and the underparts are otherwise pale gray-buff with very faint streaks on the breast and flanks. Juveniles are similar to an adult but browner and more heavily streaked.

Cassin's Sparrow is present as a breeding species across most of its North American range mainly from May to August. It is present year-round in the border regions of Texas, New Mexico, and Arizona, and its resident range extends south into Mexico. It is one of the hardest sparrows to observe, being secretive and more inclined to creep through vegetation than to fly. It is easiest to see in spring, when males sometimes sing from exposed perches.

FACT FILE

LENGTH 6 in (15 cm)

FOOD Fallen seeds and some invertebrates

HABITAT Dry grassland

STATUS Locally common summer visitor, and very local resident

VOICE Song comprises a trilling phrase preceded by, and ending with, whistling notes. Call is a thin *tseip*

Bachman's Sparrow

Peucaea aestivalis

Bachman's Sparrow is endemic to southeast U.S.A. The sexes are similar but regional variation exists. Adults from the northeast and west of the species' range have reddish-brown upperparts, the back feathers with gray margins that align to form lines. The head has a reddish brown crown with a pale central stripe, a gray supercilium and face, with a reddish-brown eye stripe and ear-covert margins. The underparts are mostly pale gray, but suffused orange-buff on the breast. Southeastern adults are similar but overall darker, with buff elements of the plumage replaced by gray. Juveniles are similar to their respective regional adults but with bolder streaking and spotting both above and below.

Bachman's Sparrow is present year-round in its southeastern U.S.A. range. It is another secretive sparrow that is more inclined to creep through vegetation than to fly. The best chances of seeing it come in spring, when males sometimes sing from exposed perches.

FACT FILE

LENGTH 6 in (15 cm)

FOOD Fallen seeds and some invertebrates

HABITAT Open grassy pine forests

STATUS Scarce resident and partial migrant

VOICE Song starts with a whistle, followed by a musical rattle. Call is a sharp *tzip*

American Tree Sparrow

Spizelloides arborea

The American Tree Sparrow is well marked by sparrow standards. The sexes are similar, and the plumage is brightest in spring and summer. Adults have a rufous back with dark streaks, and rufous wings with two whitish wingbars. The head pattern comprises a gray face, a rufous crown and a rufous stripe behind the eye, and a faint dark line bordering the gray throat. The underparts are otherwise pale gray with a rufous patch on the flanks and a dark breast spot. Juveniles are similar to an adult but heavily streaked. In all birds, the bill has a dark upper mandible and yellowish lower mandible.

The American Tree Sparrow is present as a breeding species across much of northernmost North America, mainly from April to September. Birds move south in fall and the winter range extends across most of the center of the continent. Outside the breeding season the species is typically seen in flocks.

adult

FACT FILE

LENGTH 6.25 in (16 cm)

FOOD Mainly seeds, with invertebrates in spring and summer

HABITAT Tundra in summer; rough grassland in winter

STATUS Widespread and locally common

VOICE Song is a series of shrill, warbling notes, ending with a trill. Call is a thin *tseink*

adult life-size

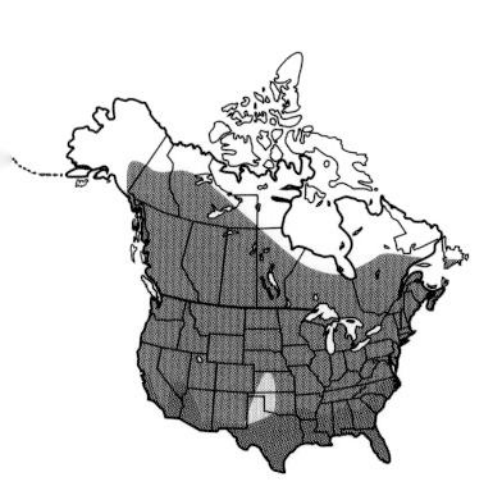

Chipping Sparrow

Spizella passerina

non-breeding adult

Thanks to its association with manmade habitats, the Chipping Sparrow is a familiar species. The sexes are similar but there is seasonal plumage variation. Breeding adults have a brown back with dark streaks, and buffish-brown wings with two whitish wingbars. The tail is dark and the pale gray rump is usually seen only in flight. The head pattern comprises a chestnut crown, a white supercilium, and a dark eye stripe. The throat is whitish and the underparts are otherwise pale gray. Non-breeding adults are similar, but the brown crown has a pale central stripe and the supercilium is paler buff. Juveniles are like a non-breeding adult but are heavily streaked; by their first winter, they resemble a pale non-breeding adult.

The Chipping Sparrow is present as a breeding species across most of central and northern North America, mainly from May to September. Most of the population migrates south in fall, boosting resident numbers in the south; its winter range extends to Central America. Outside the breeding season it often forms flocks. In town parks and gardens, birds are often tame.

FACT FILE

LENGTH 5.5 in (14 cm)

FOOD Mainly seeds, with invertebrates in spring and summer

HABITAT Open wooded habitats, including parks and gardens

STATUS Widespread and common summer visitor, partial resident, and winter visitor

VOICE Song is a rattling trill. Call is a thin *tzip*

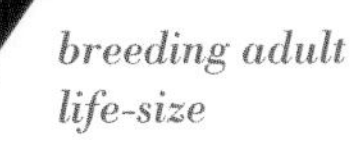

breeding adult
life-size

Clay-colored Sparrow

Spizella pallida

The Clay-colored Sparrow is a rather pale species. The sexes are similar but there is seasonal plumage variation. Breeding adults have a brown back with dark streaks, and brown wings with two white wingbars. The tail is dark and a buffish-brown rump can be seen in flight. The nape is gray and the head pattern comprises a brown crown, a whitish supercilium, dark-framed brown ear coverts, and a white "mustache" and throat. The underparts are otherwise gray-buff. Non-breeding adults are paler and more buff overall, particularly on the breast. Juveniles are similar to a non-breeding adult but heavily streaked; by their first winter, they are like a non-breeding adult but even more buff.

The Clay-colored Sparrow is present as a breeding species across northern North America mainly from May to August. It spends the rest of the year mainly in Mexico, although small numbers winter in south Texas. It forms flocks outside the breeding season.

adult life-size

FACT FILE

LENGTH 5.5 in (14 cm)

FOOD Mainly seeds, with invertebrates in spring and summer

HABITAT Scrubby grassland and prairies

STATUS Widespread and locally common summer visitor

VOICE Song is a series of insect-like buzzing trills. Call is a thin *tsip*

Brewer's Sparrow

Spizella breweri

Brewer's Sparrow is a rather nondescript little songbird with a small, pointed bill. The sexes are similar but there is seasonal and regional plumage variation. Breeding adults have a gray-buff back with dark streaks, and brown wings with two buff wingbars. Birds from the species' main range have a gray-buff nape; this feature is gray in the so-called "Timberline" Sparrow (subsp. *taverneri*, from the Pacific Northwest). In all birds, the head pattern comprises a gray-brown crown, dark-framed brown ear coverts, a pale gray supercilium and lores, and a pale throat bordered by a dark malar stripe. The underparts are otherwise pale gray-buff. Non-breeding adults are similar but paler overall and less strikingly marked. Juveniles are similar to a non-breeding adult but heavily streaked; by their first winter, they are like adult birds.

Brewer's Sparrow is present as a breeding species in western North America, mainly from April to August, with a disjunct population (the so-called "Timberline" Sparrow) found in the Pacific Northwest. Outside the breeding season, birds move south and the winter range extends across southwest U.S.A. and Mexico.

adult
life-size

FACT FILE

LENGTH 5.5 in (14 cm)

FOOD Mainly seeds, with invertebrates in spring and summer

HABITAT Sagebrush (*Artemisia* spp.) habitats and deserts in summer; grassland habitats in winter

STATUS Locally common summer visitor

VOICE Song is a series of trills and warbling whistles. Call is a thin *tsik*

Field Sparrow

Spizella pusilla

The Field Sparrow is a well-marked species with a pink bill. The sexes are similar but regional variation exists. Adults have a reddish-brown back with dark streaks, and wings with dark primaries and two white wingbars. The head has a rufous crown. The gray face is rather plain with a faint brown eye stripe in western birds, but shows brown-framed rufous ear coverts in eastern birds. The throat is pale and the underparts are otherwise gray, washed rufous on the breast and flanks, more intensely so in eastern birds than western ones. Juveniles are similar to their respective regional adults but streaked below.

The Field Sparrow is present year-round in southeast North America, but in summer (mainly from April to August) the range extends much farther north. In winter the range extends south into Mexico. Outside the breeding season the species is often seen in flocks.

FACT FILE

LENGTH 5.75 in (14.5 cm)

FOOD Mainly seeds, with invertebrates in spring and summer

HABITAT Overgrown fields with scrub

STATUS Widespread and common summer visitor, and year-round resident

VOICE Song is a series of whistling phrases that ends in a trill. Call is a sharp *tchip*

Black-chinned Sparrow

Spizella atrogularis

The Black-chinned Sparrow is a well-marked songbird with a pink bill. The sexes are subtly dissimilar in summer. Breeding adult males have a streaked reddish-brown back, and reddish-brown wings with two indistinct wingbars. The rump is a paler gray than the tail. The head has a black throat and lores, and the plumage is otherwise gray except for the white undertail. Breeding adult females are similar but the face is gray, not black. At other times, all birds are similar to an adult breeding female. Juveniles are similar to a winter adult but with faint streaking on the underparts.

The Black-chinned Sparrow is present as a breeding species in southwest U.S.A. mainly from May to August. Outside the breeding season birds move south, mainly to Mexico, although small numbers winter, or are present year-round, in border regions of Arizona, New Mexico, and Texas. The species is easiest to see in spring, when males sometimes sing from exposed perches.

FACT FILE

LENGTH 5.75 in (14.5 cm)

FOOD Mainly seeds, with invertebrates in spring and summer

HABITAT Arid, scrub-covered rocky slopes

STATUS Locally common summer visitor

VOICE Song starts with a slow, whistled *tsweet-err-tsweet*…, then speeds up and ends in a trill. Call is a thin *tsik*

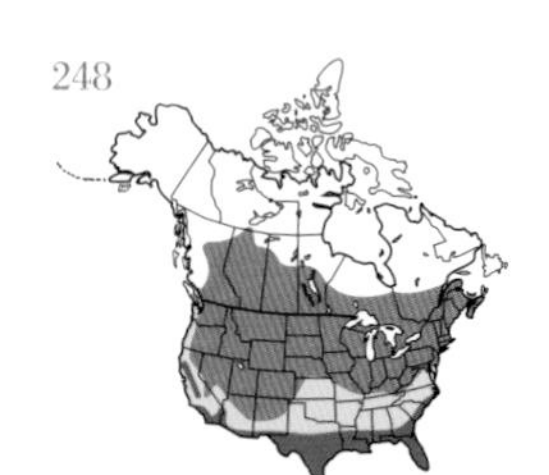

Vesper Sparrow

Pooecetes gramineus

The Vesper Sparrow is a long-tailed species with rather nondescript plumage. The sexes are similar, as are adults and juveniles. All birds have a streaked brown back, and a brown tail with white outer feathers. The wing feathers are mainly dark with pale margins, but the lesser coverts are chestnut and there are two pale wingbars. The head pattern comprises a streaked crown with a pale central stripe, a white eyering, and a dark patch on the ear coverts. The whitish throat is bordered by a dark stripe and the otherwise pale underparts are streaked on the breast and flanks.

The Vesper Sparrow is present as a breeding species across much of central North America, mainly from April to September. Birds move south outside the breeding season and the winter range extends to southern U.S.A. to Mexico. They are often found feeding beside roadsides and join flocks of mixed sparrow species outside the breeding season.

adult
life-size

adult

FACT FILE

LENGTH 6.25 in (16 cm)

FOOD Mainly seeds, with invertebrates in spring and summer

HABITAT Sagebrush (*Artemisia* spp.) habitats and dry grassland

STATUS Scarce and local summer visitor

VOICE Song (sometimes sung at dusk) starts with a few musical notes, then drawn-out whistles, and ends in a trill. Call is a sharp *tchip*

Lark Sparrow

Chondestes grammacus

The Lark Sparrow is a strikingly marked sparrow. The sexes are similar. Adults have a streaked gray-brown back, and brown wings with indistinct pale wingbars. The head pattern comprises a chestnut crown with a pale central stripe, a white supercilium and eyering, dark-framed chestnut ear coverts, and a black malar stripe that separates the white "mustache" and throat. The underparts are otherwise whitish, washed gray on the chest and with a dark central breast spot. Juveniles are similar to an adult but duller and more streaked; by their first winter they more closely resemble an adult.

The Lark Sparrow is present as a breeding species across much of central and western North America, mainly from April to August. Outside the breeding season, birds head south and the winter range extends from southern U.S.A. (where the species is present year-round in places) to Mexico. Outside the breeding season they form large flocks that feed in the open, making them easy to observe.

adult

adult life-size

FACT FILE

LENGTH 6.5 in (16.5 cm)

FOOD Mainly seeds, with invertebrates in spring and summer

HABITAT Grassland and prairies with scattered bushes

STATUS Locally common summer visitor

VOICE Song starts with two piping notes, followed by trills, buzzing phrases, and whistles. Call is a thin *tsik*

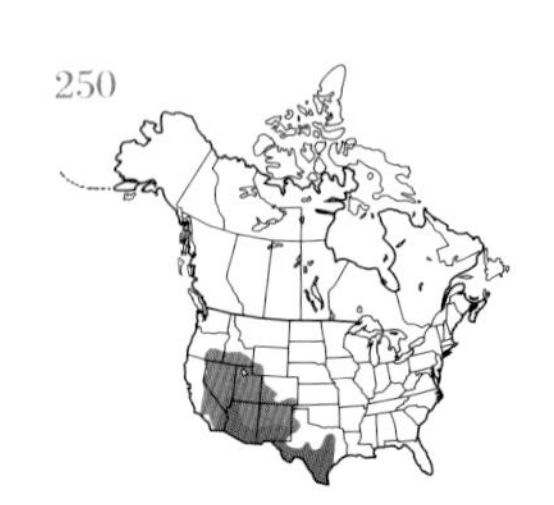

Black-throated Sparrow

Amphispiza bilineata

adult life-size

The Black-throated Sparrow is a distinctive species. The sexes are similar. Adults have mostly plain gray-brown upperparts, darkest on the wings and tail; the tail has white outer feathers that are noticeable in flight. The head pattern comprises a dark brown crown, white supercilium, dark cheeks, a white "mustache," and a black throat and chest. The underparts are otherwise whitish. Juveniles are similar to an adult but paler, with black elements of the plumage replaced by gray or streaked brown.

The Black-throated Sparrow is present as a breeding species in the north of its range mainly from May to August. Outside the breeding season these birds move south, boosting numbers of year-round residents in southern U.S.A. and Mexico. In winter, they are sometimes found in small flocks.

adult

FACT FILE

LENGTH 5.5 in (14 cm)

FOOD Mainly seeds, with invertebrates in spring and summer

HABITAT Desert and sagebrush (*Artemisia* spp.) habitats

STATUS Locally common summer visitor

VOICE Song starts with whistling notes and ends in a trill. Call is a thin, high *tink*

Sagebrush Sparrow

Artemisiospiza nevadensis

The Sagebrush Sparrow is an active arid-country species. The sexes are similar. Adults have a pale gray-brown back, rump, and tail, and brown wings with buff feather margins. The nape and crown are pale gray, and there is a white eyering and short white stripe in front of the eye. The cheeks are gray, and a gray malar stripe separates the white "mustache" from the white throat. The underparts are otherwise white, except for a central breast spot and faint streaking on the flanks. Juveniles are similar to an adult but more heavily streaked and duller.

The Sagebrush Sparrow is present as a breeding species in the north of its range mainly from May to August. Outside the breeding season, birds move south and the species' winter range extends from southern U.S.A. to northern Mexico. It is often seen running along the ground, cocking its tail up as it goes, and it is sometimes seen in small flocks outside the breeding season.

FACT FILE

LENGTH 6.25 in (16 cm)

FOOD Mainly seeds, with invertebrates in spring and summer

HABITAT Sagebrush (*Artemisia* spp.) habitats

STATUS Locally common summer visitor

VOICE Song comprises a series of grating musical phrases. Call is a thin *tsip*

adult life-size

adult

Bell's Sparrow

Artemisiospiza belli

Bell's Sparrow is an attractive little songbird. The sexes are similar. Adults have a gray-brown back, rump, and tail, and brown wings with buff feather margins. The nape, crown, and face are mainly dark gray, with a white eyering and short white line in front of the eye. A black malar stripe separates the white "mustache" from the white throat. The underparts are otherwise white except for a central black breast spot and a few dark streaks on the flanks. Juveniles are similar to an adult but duller and more heavily streaked.

adult

Bell's Sparrow was once considered to be a subspecies relative of the Sagebrush Sparrow (p.251), and formerly they were grouped together and called Sage Sparrow. Bell's Sparrow is present year-round in coastal California. Outside the breeding season it is often seen in small flocks, with members of the group flicking their tails as they feed.

adult
life-size

FACT FILE

LENGTH 6.25 in (16 cm)

FOOD Mainly seeds, with invertebrates in spring and summer

HABITAT Chaparral woodland

STATUS Locally common resident

VOICE Song comprises a series of grating musical phrases. Call is a thin *tsip*

Lark Bunting

Calamospiza melanocorys

female

The Lark Bunting is a well-marked songbird, all individuals of which have a stout conical bill and a white tip to the tail. The sexes are dissimilar. Breeding adult males have mainly black plumage but with a white patch on the wings that is conspicuous both in flight and when perched. In non-breeding adult males, black elements of the plumage are replaced by streaked and blotched gray-brown that is darker above than below. Adult females are similar to a non-breeding male but are paler overall, with a reduced amount of white on the wings. Juveniles are similar to an adult female but with less striking markings.

The Lark Bunting is present in its central North American breeding range mainly from May to August. Birds move south in fall, and the winter range extends from southern U.S.A. to Mexico. Outside the breeding season it forms large flocks.

male life-size

FACT FILE

LENGTH 7 in (18 cm)

FOOD Mainly seeds, with invertebrates in spring and summer

HABITAT Prairies in summer; rough grassland in winter

STATUS Common summer visitor; local in winter

VOICE Song is a series of whistles and trilling notes. Call is a soft *hu-eee*

Savannah Sparrow

Passerculus sandwichensis

The Savannah Sparrow shows considerable plumage variation across North America: compared to "typical" birds from the center and north of its breeding range, the so-called "Belding's Sparrow" from southern California is darkest and the so-called "Ipswich Sparrow" from the northeast is palest. Given this variation, the sexes are similar in any given location, as are adults and juveniles. All birds have brown upperparts (buff in Ipswich) with dark streaking on the back. The wings have two indistinct pale wingbars, and the inner flight feathers and greater coverts appear reddish brown in most birds. On the head, the crown has a subtle pale central stripe, and there is a yellowish supercilium and dark line behind the eye. A dark malar stripe separates the pale "mustache" and throat. The otherwise pale underparts have reddish streaks on the breast and flanks.

The Savannah Sparrow is present as a breeding species across the northern half of North America mainly from April to September. Birds migrate south in fall and the winter range for most individuals extends from southern U.S.A. to Central America; Ipswich Sparrows winter in Atlantic coastal dunes.

adult
life-size

FACT FILE

LENGTH 5.5 in (14 cm)

FOOD Mainly seeds, with invertebrates in spring and summer

HABITAT Open short grassland

STATUS Widespread and common summer visitor; locally common in winter

VOICE Song starts with a couple of *tchip* notes, followed by a trilling *bzzzrt-tzeee*. Call is a thin *stip*

Grasshopper Sparrow

Ammodramus savannarum

adult

The Grasshopper Sparrow is a well-marked little songbird. The sexes are similar. Adults have brown upperparts, with rufous margins to feathers on the back and tertials. The head pattern comprises a brown crown with a white central stripe, a buff face and lores, and a supercilium that is buff in front of the eye but grayish behind. There is a white eyering and the ear coverts are dark-framed. The underparts are whitish and plain except for the streaks and buffish wash on the breast and flanks; this feature is more obvious in fall than in spring. Juveniles are similar to an adult but heavily streaked on the breast and flanks.

The Grasshopper Sparrow is present as a breeding species mainly from April to September. Birds migrate south in fall, and the winter range extends from southern U.S.A. to Mexico. In spring males sing from exposed perches, but at other times the species is hard to observe, being generally secretive and reluctant to leave the cover of dense grasses.

FACT FILE

LENGTH 5 in (12.5 cm)

FOOD Mainly invertebrates, but also seeds

HABITAT Wide range of grassland habitats with tall vegetation

STATUS Locally common summer visitor

VOICE Song comprises a couple of *tik* notes, followed by a cricket-like trill. Call is a sharp *tsip*

adult life-size

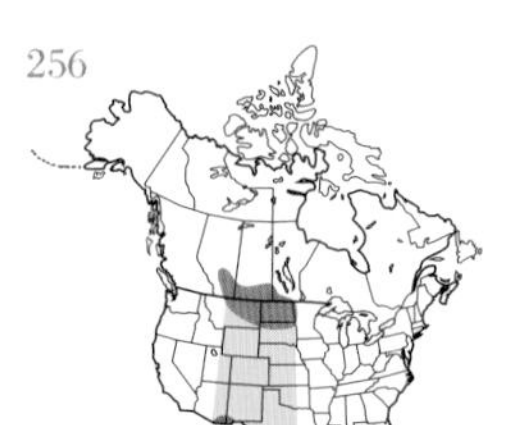

Baird's Sparrow

Ammodramus bairdii

Baird's Sparrow is a well-marked little songbird with a relatively large bill. The sexes are similar. Adults have dark brown back feathers with buff margins. The wing feathers are reddish brown with buff margins. The nape and neck are finely streaked, and the head pattern comprises a dark crown with a pale central stripe, and a buffish-yellow face with two dark spots at the rear of the ear coverts. A dark malar stripe separates the buffish "mustache" and the white throat. The underparts are otherwise pale except for the buff band across the breast and dark streaks on the breast and flanks. Juveniles are similar to an adult but the back appears scaly owing to the pale feather margins.

Baird's Sparrow is present as a breeding species in its restricted range mainly from May to September. It spends the rest of the year mainly in Mexico. It is an extremely secretive species that is most reluctant to leave the cover of grassland vegetation. The best chances for observation come in spring, when males sometimes sing from exposed perches.

FACT FILE

LENGTH 5.5 in (14 cm)

FOOD Mainly seeds, with invertebrates in spring and summer

HABITAT Shortgrass prairies in summer; short grassland in winter

STATUS Scarce and local summer visitor

VOICE Song comprises a couple of thin *tsip* notes followed by a warbled phrase and then a trill. Call is a thin *tsee*

adult life-size

Henslow's Sparrow

Ammodramus henslowii

Henslow's Sparrow is a boldly marked songbird with a proportionately large head. The sexes are similar. Adults have a reddish-brown back, the feathers with aligned pale margins. The tail and wings are reddish brown, and the wing coverts and tertials have broad rufous edges. The brown crown has a dark marginal line, and the buffish face has dark-framed ear coverts and a dark malar stripe that borders the white throat. The underparts are otherwise pale, suffused buff on the breast and with dark streaks on the breast and flanks. Juveniles are similar to an adult but paler and less strikingly marked.

adult
life-size

Henslow's Sparrow is present as a breeding species in eastern central North America, mainly from May to September. It migrates south in fall, and in winter it is found in southernmost southeast U.S.A. It is a hard species to observe, preferring to creep through vegetation rather than fly from danger. Look for it in the spring, when males sometimes sing from relatively open sites.

FACT FILE

LENGTH 5 in (12.5 cm)

FOOD Mainly invertebrates, but some seeds in winter

HABITAT Overgrown, weedy fields with matted vegetation

STATUS Local and scarce summer visitor; local and scarce in winter

VOICE Song is a short, sharp *tsi-lik*. Call is a thin *tsic*

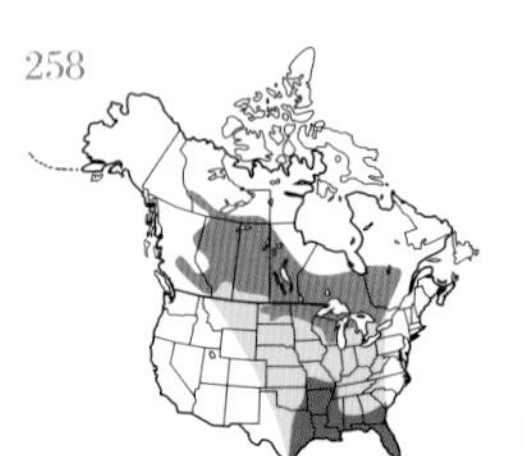

Le Conte's Sparrow

Ammodramus leconteii

Le Conte's Sparrow is a rather pale buff sparrow overall, with contrasting dark streaks. The sexes are similar. Adults have yellowish-buff upperparts, the back feathers with dark centers that align to form dark lines. The head pattern comprises a dark crown with a white central stripe, and a buffish-orange face with ear coverts that are blue-gray and show two dark spots at the rear. The underparts, including the throat, are mostly whitish with a buffish-yellow wash and faint streaks on the breast, and streaked flanks. Juveniles are similar to an adult, but the colors and markings on the head are marginally less intense.

Le Conte's Sparrow is present as a breeding species in its northern range mainly from May to September. It migrates south in fall and spends the winter months in southeast U.S.A. It is an extremely secretive species that would far rather creep away from danger through tangled vegetation than fly. Look for it in spring, when males can sometimes be seen singing.

adult

adult
life-size

FACT FILE

LENGTH 5 in (12.5 cm)

FOOD Mainly seeds, with invertebrates in spring and summer

HABITAT Wet grassland and marshes

STATUS Locally common summer visitor; local in winter

VOICE Song is a cricket-like buzzing trill. Call is a thin *tzit*

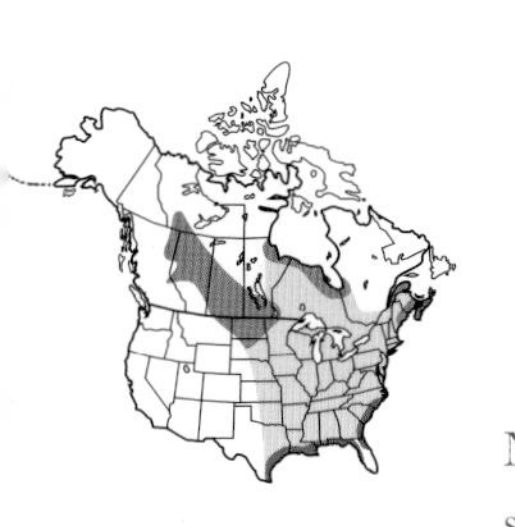

Nelson's Sparrow

Ammodramus nelsoni

Nelson's Sparrow is superficially similar to Le Conte's Sparrow (p.258), with subtle differences in plumage but entirely different habitat preferences. Birds that breed in the interior are subtly more colorful than those from the east of the species' breeding range or from the Arctic. Given these regional differences, the sexes are similar. Adults have a buffish-brown back and wings, the back feathers with dark centers and pale margins that align to form lines. The head pattern comprises a gray-centered dark crown, and a buffish-yellow face with a dark line behind the eye and blue-gray ear coverts; the nape is also blue-gray. The throat is pale buff, and the breast and flanks are streaked and washed with yellowish buff; the underparts are otherwise white. Juveniles are similar to an adult but with more colorful upperparts and much-reduced streaking on the underparts.

adult
life-size

adult

Nelson's Sparrow is present as a breeding species, mainly from June to September, in freshwater wetlands in the interior of its range but in saltmarshes elsewhere. In fall, all birds migrate to saltmarsh habitats on the Atlantic coast. It is a very secretive songbird that runs and creeps from danger, and seldom willingly takes flight.

FACT FILE

LENGTH 4.75 in (12 cm)

FOOD Mainly seeds, with invertebrates in spring and summer

HABITAT Freshwater wetlands and saltmarshes in summer; coastal saltmarshes in winter

STATUS Locally common summer visitor; local in winter

VOICE Song is a wheezy trill, delivered in a short vertical display flight. Call is a sharp *tsic*

Saltmarsh Sparrow

Ammodramus caudacutus

The Saltmarsh Sparrow is similar to Nelson's Sparrow (p.259) and favors similar habitats, especially in winter. Its longer bill and subtle plumage differences help with identification. The sexes are similar. Adults have brown upperparts, the back feathers with dark centers and white margins that align to form lines. The head pattern comprises a gray-centered dark crown and a buffish-yellow face with a dark line behind the eye and blue-gray ear coverts; there is subtle streaking on the supercilium behind the eye (the supercilium is unmarked in Nelson's). The nape is blue-gray. The throat is pale buff, and the breast and flanks are washed with yellowish buff and adorned with dark streaks that are bolder than in Nelson's. The underparts are otherwise white.

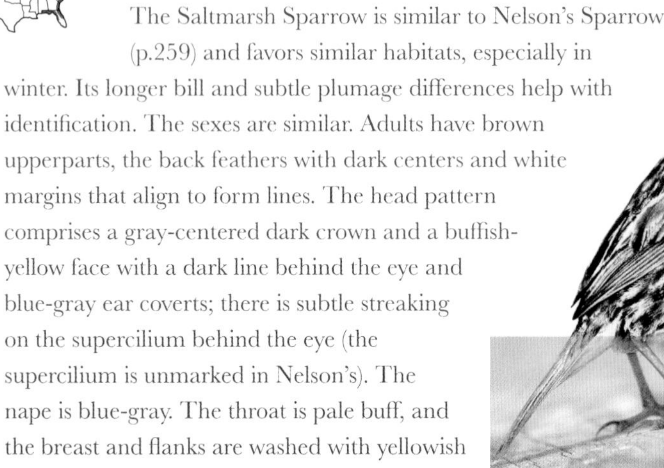

adult

The Saltmarsh Sparrow is found only on the Atlantic coast, with its summer range reaching north to Maine and its winter range extending south to Florida. It is another very secretive sparrow that is hard to observe as it creeps through vegetation.

adult
life-size

FACT FILE

LENGTH 5 in (12.5 cm)

FOOD Mainly seeds, with invertebrates in spring and summer

HABITAT Saltmarsh habitats, both in summer and winter

STATUS Locally common summer visitor; local in winter

VOICE Song is a hissing trill, delivered in a low, fluttering display flight. Call is a sharp *tsic*

Seaside Sparrow

Ammodramus maritimus

adult

The Seaside Sparrow has a relatively long, pointed bill and shows regional variation in its plumage. Three groups are recognized and the sexes are similar in any given location. "Atlantic Coast" adults have gray-brown upperparts with rufous on the wings. On the head, a dark malar stripe separates the white throat from the faintly yellow "mustache." There is a yellow supercilium in front of the eye. The underparts are dull gray with reddish-brown spots. "Gulf Coast" adults are similar but overall more intensely washed yellow-buff, and with bolder dark markings on the head, including a complete supercilium, and more intense streaks on the back and breast. "Cape Sable" (Florida) adults have more evenly brown upperparts and wings, and whitish underparts with dark brown streaks. Juveniles are similar to their respective regional adults but warmer buff overall.

The Seaside Sparrow has specialized habitat requirements and is present year-round in saltmarshes up and down North America's Atlantic and Gulf coasts. As sparrows go, it is fairly bold and easy to see. It is threatened by habitat loss and degradation.

adult
life-size

FACT FILE

LENGTH 6 in (15 cm)

FOOD Mainly seeds, with invertebrates in spring and summer

HABITAT Grassy saltmarshes

STATUS Very locally common resident

VOICE Song is a wheezy *tsup-bree-erz*. Call is a soft *tsup*

Fox Sparrow

Passerella iliaca

The Fox Sparrow is a thickset sparrow whose breast markings usually consolidate to form a dark central patch. The plumage and bill size vary across its extensive breeding range, with four subspecies "groups" being recognized: "Red," "Sooty," "Slate-colored," and "Thick-billed." In any given region the sexes are similar, as are adults and juveniles. "Reds" have a gray and reddish back and wings, two faint wingbars, and a red and gray face pattern. The underparts are boldly streaked reddish brown. "Sooty" birds are similar but have darker, more uniformly sooty-brown upperparts and no wingbars. In "Slate-colored" birds, the back and hood are slate-gray. In "Thick-billed" birds, the bill is large and stout.

The Fox Sparrow is present across its breeding range mainly from May to August. "Red" birds are widespread across the Arctic; "Sooty" birds breed in the Pacific Northwest; "Slate-colored" birds are found in the Rocky Mountains; and "Thick-billed" birds breed in west coast mountains. Birds migrate south in fall, and the winter range extends from southern and western U.S.A. to Mexico.

"Red" adult life-size

FACT FILE

LENGTH 7 in (18 cm)

FOOD Mainly seeds, with invertebrates in spring and summer

HABITAT Northern and montane low, scrubby woodland

STATUS Widespread and common summer visitor and winter visitor, according to region

VOICE Song comprises various whistling phrases, such as *twee-su-swee*. Call is a sharp *tchiup*

"Sooty" adult

Song Sparrow

Melospiza melodia

The Song Sparrow has breast markings that usually coalesce to form a central spot. The plumage varies across the species' range but in any given area the sexes are similar. Eastern birds have a streaked brown back, reddish-brown wings with two pale wingbars, and a reddish tail. The head has a brown crown with a pale central stripe, a gray-brown face, and a dark stripe behind the eye that emphasizes the pale supercilium. A dark malar stripe separates the pale "mustache" from the whitish throat. The underparts are otherwise pale but heavily streaked on the breast and flanks. Birds from California have cleaner-looking underparts with more strongly contrasting dark markings, and grayer-brown upperparts with contrasting reddish wings. Birds from the southwest are paler and more buff overall. Pacific Northwest birds are darker overall. Alaskan birds are appreciably larger, and duller and darker. Juveniles are similar to their respective regional adults but more buff overall.

eastern adult

The Song Sparrow is present year-round in much of central North America. Northern populations migrate south in fall, and in winter the species' range extends to southernmost U.S.A. and the Mexican border.

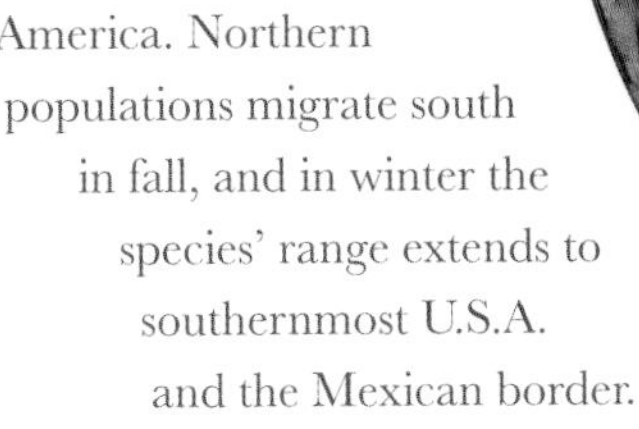

FACT FILE

LENGTH 6–7 in (15–18 cm)

FOOD Mainly seeds, with invertebrates in spring and summer

HABITAT Wide range of scrubby habitats

STATUS Widespread and common summer visitor, resident, and winter visitor, according to region

VOICE Song comprises three or four whistles followed by rich phrases and a trill. Call is a flat *cheerp*

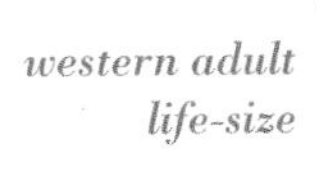

western adult
life-size

Lincoln's Sparrow

Melospiza lincolnii

Lincoln's Sparrow is a rather dumpy, short-tailed species. The sexes are similar. Adults have a dark-streaked buffish-brown back, and reddish-brown wings with two faint wingbars; the tail is reddish brown. The head has a brown crown with a pale central stripe, and dark-framed ear coverts that help define the broad gray supercilium. There is a buff malar stripe and a dark line that borders the streaked whitish throat. The breast and flanks are streaked and suffused buff, and the underparts are otherwise whitish. Juveniles are similar to an adult but paler overall, with more distinct wingbars.

Lincoln's Sparrow is present as a breeding species across the Arctic and in the Rockies, mainly from April to September. Birds migrate south in fall, and the winter range extends from southern and western U.S.A. throughout Central America. It is a relatively bold species and usually easy to see in suitable habitats.

adult life-size

adult

FACT FILE

LENGTH 5.75 in (14.5 cm)

FOOD Mainly seeds, with invertebrates in spring and summer

HABITAT Overgrown and weedy fields

STATUS Widespread and common summer visitor

VOICE Song starts with a series of warbling jingles and ends in a trill. Call is a soft *tchup*

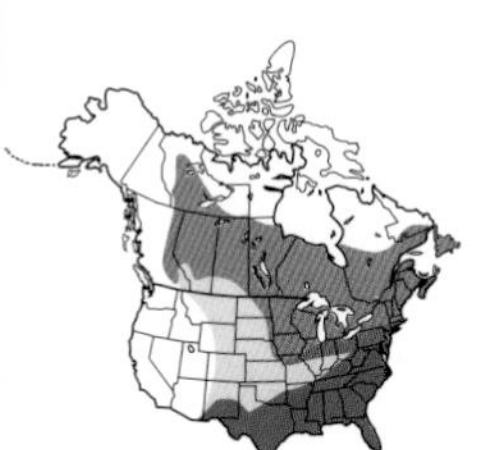

Swamp Sparrow

Melospiza georgiana

The Swamp Sparrow is a well-marked, dumpy little songbird. The sexes are similar. Summer adults have a streaked brown back, and rufous wings and tail. The head has a rufous crown and dark-framed gray ear coverts that help emphasize the gray supercilium and buffish malar stripe. The white throat is bordered by a dark line, and the underparts are otherwise mostly plain gray, faintly streaked on the breast and suffused buff on the flanks. Winter adults and immatures are similar to a summer adult, but the colors are much duller and the markings are less striking. Juveniles are similar to a winter adult but browner overall and more heavily streaked.

adult

The Swamp Sparrow is present as a breeding species across much of northern North America (except the far northwest) mainly from April to September. It migrates south in fall, and the winter range extends from southeast U.S.A. to Central America.

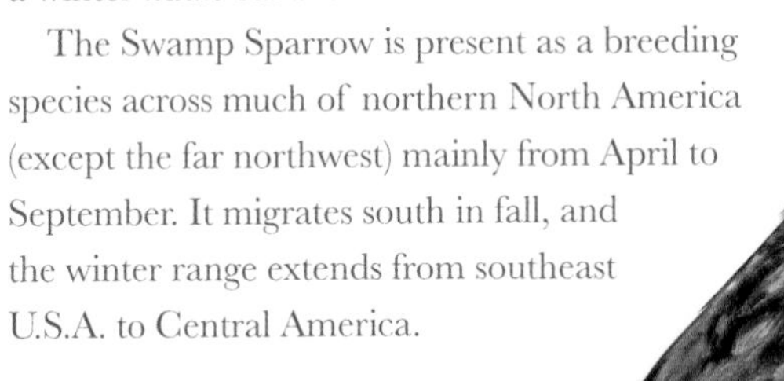

FACT FILE

LENGTH 5.75 in (14.5 cm)

FOOD Mainly seeds, with invertebrates in spring and summer

HABITAT Marshes, inundated scrubby woodlands, and swamp margins

STATUS Widespread and common summer visitor; widespread in winter

VOICE Song is a short trilling rattle. Call is a sharp *tchip*

adult life-size

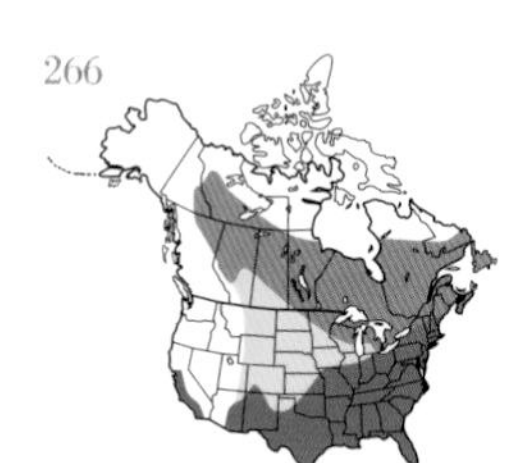

White-throated Sparrow

Zonotrichia albicollis

The White-throated Sparrow is a distinctive woodland songbird. The sexes are similar but two color forms occur. All adults have a dark-streaked brown back and reddish-brown wings with two white wingbars. The tail is gray-brown. In typical "White-striped" birds the head has a dark crown with a pale central stripe, and a broad supercilium that is yellow-buff in front of the eye but white behind. In "Tan-striped" birds the supercilium is uniformly yellow-buff, and the central crown stripe is buffish gray. All birds have a black eye stripe, gray cheeks, a white throat, a gray breast, and otherwise whitish underparts. Juveniles recall a heavily streaked adult with indistinct head markings; by their first winter they recall a dull "Tan-striped" adult.

The White-throated Sparrow is present as a breeding species in northern forests mainly from April to August. Birds migrate south in fall, and the winter range is mainly southeast U.S.A. and also down the Pacific coast. Small numbers can be found year-round in the northeast. The species forms flocks outside the breeding season and visits feeders in winter.

"White-striped" adult life-size

FACT FILE

LENGTH 6.75 in (17 cm)

FOOD Mainly seeds, with invertebrates in spring and summer

HABITAT Northern forests in summer; dense woodland and scrub in winter

STATUS Widespread and common, both in summer and winter

VOICE Song is a piercing, whistling *see-tsee-chrrdede-chrrdede*. Call is a sharp *chink*

"Tan-striped" adult

Harris's Sparrow

Zonotrichia querula

Harris's Sparrow is a plump, boldly marked bird with a distinctive pink bill. The sexes are similar. Summer adults have a dark-streaked buffish-brown back and brown wings with two white wingbars. The face is mainly gray and the head is otherwise adorned with a black crown, ear covert patch, face, throat, and bib. Typically, the bib is larger in males than females and gets more extensive with age. The underparts are otherwise whitish with dark streaks on the flanks. Winter adults are similar but the face is buffish brown, not gray, and the crown is speckled. Juveniles are similar but have no black on the face, and the breast and flanks are streaked brown. By their first winter, they have acquired a hint of an adult's black face markings, although the throat is white and the breast is streaked.

Harris's Sparrow is restricted as a breeding species to central northernmost Canada, where it is present mainly from May to September. It migrates south in fall and in winter it occurs in the Great Plains region. Outside the breeding season it forms flocks and often mixes with other sparrow species.

FACT FILE

LENGTH 7.5 in (19 cm)

FOOD Mainly seeds, with invertebrates in spring and summer

HABITAT Stunted boreal forests in summer; scrubby woodland and brush in winter

STATUS Very locally common, both in summer and winter

VOICE Song is a series of penetrating whistles. Call is a sharp *twink*

adult

1st-winter life-size

White-crowned Sparrow

Zonotrichia leucophrys

dark-lored adult

The White-crowned Sparrow is a distinctive songbird. The sexes are similar. Adults have a dark-streaked brown back and reddish-brown wings with two white wingbars. The tail and rump are gray-brown. The head pattern comprises a black eye stripe, and a black crown with a broad white central stripe. In eastern and Rocky Mountain birds, the bill is pink, the lores are black, and there is a white supercilium behind the eye; the underparts are otherwise mostly gray and palest on the belly. In otherwise similar western and northwestern birds, the lores are gray and the white supercilium is complete. The bill is orange in western birds but yellow in northwestern individuals. Juveniles recall their respective regional adults but are heavily streaked; by their first winter, their plumage is closer to that of an adult but black elements of the head markings are brown.

The White-crowned Sparrow is present as a breeding species across Arctic North America and in western mountain ranges. Birds migrate south in fall, and the winter range extends across southern U.S.A. and into Mexico. The species forms flocks outside the breeding season.

pale-lored adult
life-size

FACT FILE

LENGTH 7 in (18 cm)

FOOD Mainly seeds, with invertebrates in spring and summer

HABITAT Taiga woodland in summer; wide range of wooded habitats in winter

STATUS Widespread and common, both in summer and winter

VOICE Song comprises a couple of shrill whistles followed by several squeaky chirps. Call is a sharp *pink*

Golden-crowned Sparrow

Zonotrichia atricapilla

winter adult

The Golden-crowned Sparrow is distinctive as a summer adult, but less so at other times. The sexes are similar. Summer adults have a dark-streaked brown back and reddish-brown wings with two white wingbars. The rump and tail are gray-brown. The nape is gray, and the head pattern comprises a gray face and a black crown with a broad yellow central stripe. The underparts are brown on the breast and flanks but otherwise pale gray, palest on the undertail. Winter adults are similar, but dark elements on the crown are speckled and the central stripe is less colorful. Juveniles are similar to a winter adult but the head pattern is indistinct, with only a hint of color, and the underparts are heavily streaked. By their first winter, their plumage is closer to a winter adult's but the head markings and colors remain indistinct.

The Golden-crowned Sparrow is a western specialty that is present as a breeding species in the Pacific Northwest mainly from May to August. In winter, its range extends down the west coast to the Mexican border. Outside the breeding season it forms flocks.

breeding adult life-size

FACT FILE

LENGTH 7 in (18 cm)

FOOD Mainly seeds, with invertebrates in spring and summer

HABITAT Wet boreal forests in summer; dense woodland and chaparral in winter

STATUS Locally common, both in summer and winter

VOICE Song is a series of whistled notes such as *see-er-duu-see*. Call is a soft *tseek*

Dark-eyed Junco

Junco hyemalis

Distribution of "Slate-colored" group

The Dark-eyed Junco has a huge geographical range, with regional plumage differences to match. The large number of subspecies are clumped together in plumage "groups," the commonest of which are covered here. Adult male "Slate-colored" birds (widespread in the east) have slate-gray plumage except for the white belly and undertail. Adult male "Oregon" birds (widespread in the west) have a black hood, a reddish-brown back and gray rump, dark wings, and white underparts with a reddish wash on the flanks. Adult male "Gray-headed" birds (from the southern Rockies) have a gray hood with black lores, a reddish back and gray rump, and a dark tail and wings; the underparts are pale gray. Adult females recall their respective regional males but their upperparts are usually browner overall. Juveniles are similar to a respective regional adult female, but browner still and streaked on the underparts.

"Gray-headed" male

"Slate-colored" male life-size

FACT FILE

LENGTH 6.25 in (16 cm)

FOOD Mainly seeds, with invertebrates in spring and summer

HABITAT Northern and boreal forests in summer; a wide range of woodlands in winter

STATUS Widespread and common, both in summer and winter

VOICE Song is a rapid trill. Call is a sharp *tchht*

The Dark-eyed Junco is present as a summer breeding visitor to northern North America, mainly from May to August. These populations migrate south in fall. Birds are present year-round in the west and northeast, and the species' winter range extends south across the continent. Outside the breeding season it often forms flocks.

"Pink-sided" male; this form is found in the northern Rockies

"Oregon" male

"Oregon" female

Hepatic Tanager

Piranga flava

The Hepatic Tanager has colorful but rather uniform plumage. The sexes are dissimilar. Adult males are red overall, brightest on the crown, throat, and undertail, and with a gray tint to the ear coverts and back. Females of all ages, and immature males, are yellow overall, brightest on the crown, throat, and undertail, and with a gray tint to the ear coverts and back. Juveniles are similar to an adult female but with faint streaking on the back.

male

female

The Hepatic Tanager is present as a breeding species in southwest U.S.A. mainly from May to August. It spends the rest of the year in Mexico, where the species also breeds and is resident. It is not a particularly shy bird but can be hard to spot as its plumage colors blend in surprisingly well in dappled foliage.

FACT FILE

LENGTH 8 in (20 cm)

FOOD Invertebrates in spring and summer; fruits and berries at other times

HABITAT Upland pine and mixed forests

STATUS Locally common summer visitor

VOICE Song is a series of short musical, two-note whistles. Call is a soft *tchuk*

male
life-size

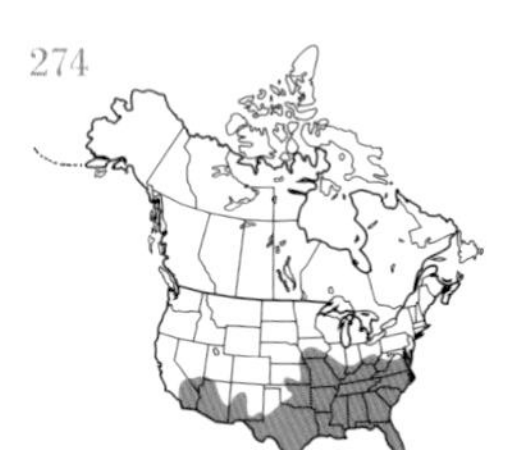

Summer Tanager

Piranga rubra

The Summer Tanager is a colorful and attractive songbird with a peaked, not rounded, crown. The sexes are dissimilar. Adult males are bright red, but subtly darker on the wings and tail than elsewhere. Most females of all ages, and most immature males, are yellow-buff, but subtly darker on the wings, back, and tail than elsewhere; some individuals are very subtly mottled with red. By their first spring, males are blotchy red on the head, neck, breast, and back.

FACT FILE

LENGTH 7.75 in (19.5 cm)

FOOD Invertebrates, fruits, and berries

HABITAT Deciduous and mixed woodland

STATUS Widespread and common summer visitor

VOICE Song is a series of warbling whistles, recalling that of an American Robin (p.140). Call is a rattling *pi-tuk*

male
life-size

The Summer Tanager is present as a breeding species in southern U.S.A. mainly from May to August. It spends the rest of the year in Central America. The species is easily overlooked when perched or feeding in dappled foliage.

Scarlet Tanager

Piranga olivacea

A breeding male Scarlet Tanager is one of North America's most stunningly colorful songbirds. The sexes are dissimilar. Summer adult males have bright red body plumage with a black tail and wings. The bill color ranges from pink to gray. Adult males in fall, and immature males, resemble a breeding male but red elements of the plumage are greenish yellow. At all times, females have greenish-yellow plumage with a dark tail and wings (not quite as dark as in a fall male).

male

female

The Scarlet Tanager is present as a breeding species in eastern North America mainly from May to August. It spends the rest of the year in South America. Despite its colorful plumage the species can be hard to spot among dappled tree foliage, so listen for its song to detect its presence.

male
life-size

FACT FILE

LENGTH 7 in (18 cm)

FOOD Invertebrates, fruits, and berries

HABITAT Deciduous woodland

STATUS Widespread and common summer visitor

VOICE Song is a series whistled phrases whose tone recalls that of an American Robin (p.140). Call is a sharp *tchh-brrr*

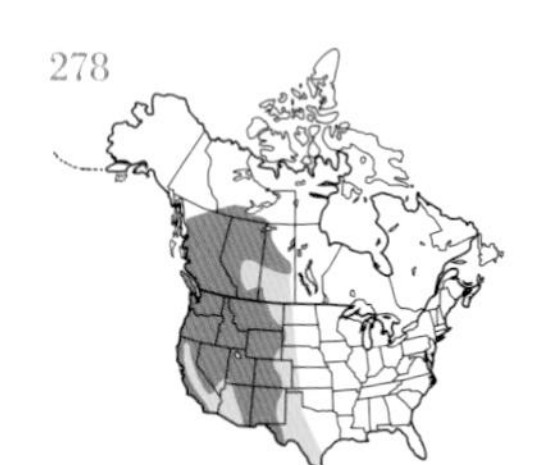

Western Tanager

Piranga ludoviciana

A male Western Tanager is colorful and almost unmistakable. The sexes are dissimilar. Adult summer males have a black back, tail, and wings, the latter with two wingbars, the upper of these yellow and the lower white. The body plumage is otherwise yellow, and the head is flushed with red. Adult winter males are similar, but yellow elements to the plumage are grubbier and most of the red color is lost, with just a hint retained at the base of the bill. Adult females are similar to a winter male, but black elements of the plumage are gray-green, the red color is entirely absent, and the wingbars are less distinct. Immatures resemble their respective winter adults but are paler and less colorful overall.

male
life-size

FACT FILE

LENGTH 7.25 in (18.5 cm)

FOOD Invertebrates, fruits, and berries

HABITAT Conifer and mixed woodland

STATUS Widespread and common summer visitor

VOICE Song is a series of fluty two-note whistles. Call is a rattling *trrrt*

The Western Tanager is present as a breeding species in western North America mainly from May to August. It spends the rest of the year in Central America. Like other tanagers, its colorful plumage helps it blend in remarkably well with dappled foliage in the tree canopy.

Northern Cardinal

Cardinalis cardinalis

With its peaked crest, the Northern Cardinal is an unmistakable and iconic North American songbird. The sexes are dissimilar. Adult males have a black face but otherwise bright red plumage and a bright red bill. Adult females have gray-buff body plumage with a subtly darker face, a red tail, red tip to the crest, and a red tint to the wings; the bill is dull red. Juveniles are similar to an adult female but the plumage is dull gray-buff overall, with a reddish flush to the breast and tail that is particularly noticeable in males; all juveniles have a dark bill. In flight, all birds reveal red underwing coverts that are brightest in adult males, dullest in juveniles.

The Northern Cardinal is present year-round across most of eastern North America and parts of the southwest. In many locations it is accustomed to people and often visits garden feeders in winter.

FACT FILE

LENGTH 8.75 in (22 cm)

FOOD Mainly seeds and fruits

HABITAT Wide range of wooded habitats, including parks and gardens

STATUS Widespread and common resident

VOICE Song is a series of fluty whistles, typically either *tiu-tiu-tiu-tiu* or *ptee-ptee-ptee*. Call is a sharp *tik*

male life-size

Pyrrhuloxia

Cardinalis sinuatus

The Pyrrhuloxia is the southwestern desert counterpart of the Northern Cardinal (p.280), and has distinctive plumage and a thickset parrot-like bill. The sexes are dissimilar. Adult males have a gray back, subtly darker wings with a red leading edge, and a mostly red tail. The plumage is otherwise mainly pale gray except for the red crest, and red that extends from the face to the center of the breast and belly. Adult females are mainly buffish gray, flushed pink on the center of the breast and belly; the tail has red outer feathers, and the leading edge to the wings and tip of the crest are red. Juveniles are similar to an adult female but duller and less colorful overall; the bill is dark. In flight, all birds reveal red underwing coverts that are brightest in males, dullest in juveniles.

The Pyrrhuloxia is present year-round in arid southwestern habitats from Texas to southern Arizona. It is sometimes seen in roving flocks outside the breeding season and occasionally visits feeders.

FACT FILE

LENGTH 8.75 in (22 cm)

FOOD Mainly seeds and fruits

HABITAT Scrubby desert habitats

STATUS Locally common resident

VOICE Song is a series of subdued whistles. Call is a sharp *chint*

female

male
male
life-size

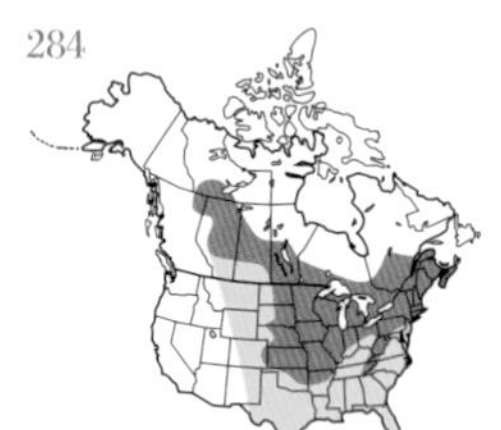

Rose-breasted Grosbeak

Pheucticus ludovicianus

The Rose-breasted Grosbeak is a well-marked songbird with a large bill. The sexes are dissimilar. Summer adult males have a black hood and back, and black wings with white patches and bars. The breast is pinkish red and the underparts are otherwise white, as is the rump. In flight, red underwing coverts are revealed. Adult males in fall are similar, but black elements of the plumage are streaked brown. Adult females and juveniles have streaked brown plumage that is paler below than above. The wings have two white wingbars and the head has a pale supercilium. By fall, immature males are flushed pinkish orange on the breast, and their wingbars are broader than those in an immature female.

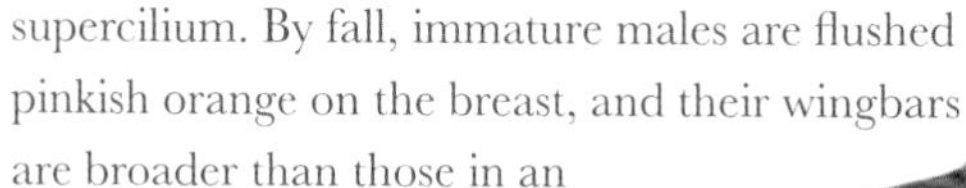

immature male

FACT FILE

LENGTH 8 in (20 cm)

FOOD Mainly invertebrates in spring and summer; fruits and berries at other times

HABITAT Open deciduous woodland

STATUS Widespread and common summer visitor

VOICE Song is a series of fluty whistles, similar to that of an American Robin (p.140). Call is a sharp *peek*

The Rose-breasted Grosbeak is present as a breeding species in northern wooded regions mainly from May to August. It spends the rest of the year in Central America. Silent birds are rather unobtrusive and easy to overlook in dappled foliage.

Black-headed Grosbeak

Pheucticus melanocephalus

The Black-headed Grosbeak is large-billed songbird with similar proportions to the Rose-breasted Grosbeak (p.284). The sexes are dissimilar. Adult males have a streaked, dark brown back, dark wings with white wingbars, and a dark tail that contrasts with the buffish-orange rump. The head has a dark hood; the lower throat and rest of the underparts are orange-buff. Adult females have streaked brown upperparts and dark wings with two white wingbars. The head has a brown cap with a white supercilium. The throat is pale, but the underparts are otherwise pale orange-buff with faint streaking on the flanks. Juveniles are similar to an adult female but more heavily streaked. By their first winter immatures are similar to an adult female, but in males the underparts are tinged orange-buff more intensely than in females. In all birds, the underwing coverts are yellow.

male

female

The Black-headed Grosbeak is present as a breeding species in western North America mainly from May to August. It spends the rest of the year in Mexico.

male life-size

FACT FILE

LENGTH 8.25 in (21 cm)

FOOD Mainly invertebrates in spring and summer; fruits and berries at other times

HABITAT Conifer forests

STATUS Widespread and common summer visitor

VOICE Song comprises warbling and whistling phrases, similar to those of both the American Robin (p.140) and Hepatic Tanager (p.272). Call is a sharp *eek*

Blue Grosbeak

Passerina caerulea

The Blue Grosbeak is an attractive songbird with a relatively large, conical bill. The sexes are dissimilar. Adult males are mostly deep blue, with a black face, two reddish-brown wingbars, and buff margins to the tertial feathers. Adult females are mainly buffish brown, subtly darker above than below, and with a pale throat and two reddish-brown wingbars. Juveniles are similar to an adult female. By their first spring, immature males have acquired a hint of blue on the head, rump, and tail.

The Blue Grosbeak is present as a breeding species across the southern half of North America mainly from May to August. It spends the rest of the year mainly in Central America. Typically, it remains hidden in dense brush for much of the time, but males in particular sometimes perch in the open, fanning their tail in an agitated manner.

male life-size

female

1st-spring male

FACT FILE

LENGTH 6.75 in (17 cm)

FOOD Invertebrates and seeds in spring and summer; fruits and berries at other times

HABITAT Scrub-colonized overgrown grassland

STATUS Widespread and common summer visitor

VOICE Song is a series of warbling whistles. Call is a sharp *pink*

Lazuli Bunting

Passerina amoena

The Lazuli Bunting is a colorful songbird and the western counterpart of the Indigo Bunting (p.290). The sexes are dissimilar. Summer adult males have a blue hood and back, and dark gray wings, tinged blue and with a white wingbar. The breast is orange-buff and the underparts are otherwise white. Winter adult males are similar, but blue elements of the plumage are muted by brown feather margins and the plumage is less colorful overall. Adult females have a buffish-brown hood and back, a bluish rump that contrasts with the darker tail, and dark wings with two indistinct pale wingbars. The breast is washed buff and the underparts are otherwise whitish. Juveniles are similar to an adult female but with more buff plumage overall; by spring, immature males have acquired some blue coloration on the head and upperparts.

The Lazuli Bunting is present as a breeding species in western North America mainly from May to August. It spends the rest of the year mainly in Mexico, and forms sizeable flocks outside the breeding season.

FACT FILE

LENGTH 5.5 in (14 cm)

FOOD Invertebrates in spring and summer; mainly seeds at other times of year

HABITAT Open deciduous woodland and scrub

STATUS Locally common summer visitor

VOICE Song is a series of tuneful whistling phrases. Call is a sharp *tchht*

Indigo Bunting

Passerina cyanea

The Indigo Bunting is a familiar little songbird, and the eastern counterpart of the Lazuli Bunting (p.289). The sexes are dissimilar. Summer adult males are blue, darkest on the head and grayest on the wings. The conical bill is pale gray. In winter adult males, the plumage colors are masked by brown feather margins; these wear away by spring to reveal blue plumage. Adult females are brown, darker above than below, with faint streaking on the underparts and two indistinct wingbars. Juveniles are similar to an adult female; by their first spring, immature males have acquired some blue feathers but look very blotchy overall.

The Indigo Bunting is present as a breeding species across the eastern half of North America mainly from May to September. It spends the rest of the year in Central America and the Caribbean region, with small numbers wintering in southern Florida. Perched birds often twitch their tail in an agitated manner. Outside the breeding season it forms flocks.

FACT FILE

LENGTH 5.5 in (14 cm)

FOOD Mainly seeds, with invertebrates in spring and summer

HABITAT Scrub, deciduous woodland, and overgrown fields

STATUS Widespread and common summer visitor

VOICE Song is a series of whistling chirps that ends in a trill. Call is a sharp *stik*

Painted Bunting

Passerina ciris

The Painted Bunting is a striking little songbird. The sexes are dissimilar. Adult males have a blue hood with a narrow red line down the throat, the color continuing on the underparts and rump. The back is yellowish green and the brown wings have green feather margins. Adult females are plain by comparison, with yellowish-green upperparts and paler yellow underparts. Juveniles recall an adult female, but their overall buff plumage shows just a tinge of green on the upperparts; by their first spring, immature males have acquired some of the adult male's blue and red colors.

The Painted Bunting is present as a breeding species in southeast U.S.A. mainly from May to September. It spends the rest of the year mainly in Central America, although small numbers winter in southern Florida. Despite the male's gaudy colors, the species' secretive nature and habit of keeping to dense cover make it hard to see.

female

immature male

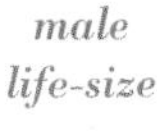

male life-size

FACT FILE

LENGTH 5.5 in (14 cm)

FOOD Mainly seeds, with invertebrates in spring and summer

HABITAT Dense woodland margins and clearings, often near water

STATUS Locally common summer visitor

VOICE Song is a series of whistling phrases. Call is a sharp *chip*

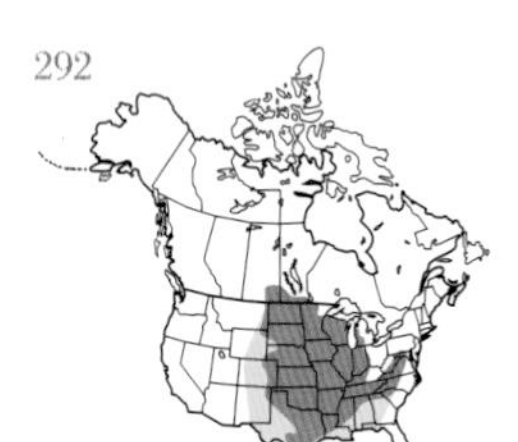

Dickcissel

Spiza americana

The Dickcissel is a well-marked sparrow-like songbird. The sexes are dissimilar. Summer adult males have a dark-striped gray-brown back, reddish-brown wings, and a gray tail and nape. The head is gray overall, but with a striking yellow supercilium and eye surround, a yellow malar stripe, and a black bib surrounding a white throat patch. The underparts are flushed yellow on the breast, grading to grayish white on the belly; the undertail coverts are white. Winter adult males have more subdued plumage coloration and the black bib is obscured by pale feather fringes. Adult females recall a dull, washed-out male with no black bib and a more extensive white throat. Juveniles recall a plain adult female with a hint of an adult's face pattern.

FACT FILE

LENGTH 6.25 in (16 cm)

FOOD Mainly seeds, with invertebrates in spring and summer

HABITAT Prairies and agricultural grassland

STATUS Common summer visitor

VOICE Song is a repeated, vaguely onomatopoeic *dik-dik-dikcissel*. Call is a buzzing *bzzrtt*

male
life-size

The Dickcissel is present as a breeding species in the Midwest of North America mainly from May to August. It spends the rest of the year mainly in the Llanos grasslands of northern Venezuela, forming huge flocks outside the breeding season.

male life-size

female

Bobolink

Dolichonyx oryzivorus

female

male

The Bobolink is an open-country songbird, the males of which are particularly striking. The sexes are dissimilar. Summer adult males are largely black, with a buff nape and white rump. The wings have a white "shoulder" patch, and white margins to the tertials and greater coverts. In all other plumages, birds are buffish brown with dark streaking on the back, and wing covert and tertial feathers showing dark centers with buff margins. The head has a dark crown with a pale central stripe, and a dark stripe behind the eye. The throat is pale and the underparts are otherwise pale buff, with dark streaks on the flanks.

The Bobolink is present as a breeding species across central North America mainly from May to August. It spends the rest of the year in Central America. The male performs a song flight in spring. Outside the breeding season the species forms flocks.

FACT FILE

LENGTH 7 in (18 cm)

FOOD Mainly seeds, with invertebrates in spring and summer

HABITAT Meadows and tallgrass prairies

STATUS Widespread and common summer visitor

VOICE Song (given in flight) is a series of fluty *bob-o-liink* phrases followed by chattering notes. Call is a harsh *chink*

male
life-size

Red-winged Blackbird

Agelaius phoeniceus

The Red-winged Blackbird is a familiar and distinctive songbird. The sexes are dissimilar. Adult males are largely black but have a bold red "shoulder" patch that is edged with yellow in some populations. In winter, the black feathers have subtle brown margins that wear off with time. Adult females are gray-brown and heavily streaked; the plumage is palest (and usually tinged pinkish buff) on the throat, and the head has a pale supercilium. Immature males are similar to a winter adult male but with more extensive brown edging to the feathers. Immature females recall an adult female but lack the pinkish-buff tinge to the throat.

female

displaying male

FACT FILE

LENGTH 8.75 in (22 cm)

FOOD Mainly seeds, with invertebrates in spring and summer

HABITAT Farmland and marshy grassland

STATUS Widespread and very common, in the north in summer but present year-round in the south

VOICE Song is a series of harsh screeches. Call is a sharp *tchik*

The Red-winged Blackbird is present year-round across the southern half of North America. In spring, the breeding range extends north to the edge of the Arctic, and the winter range extends south to Central America. Outside the breeding season the species forms huge flocks.

male
life-size

Tricolored Blackbird

Agelaius tricolor

The Tricolored Blackbird is similar to the Red-winged Blackbird (p.296). Males are separable with care but females are problematic. Adult males are largely black with a red "shoulder" patch that is edged with white (not yellow, as seen in some Red-winged Blackbird males). In winter, the black feathers have gray-buff margins. All females are dark gray-brown and heavily streaked. The plumage is palest on the throat, malar stripe, and supercilium, and lacks the pinkish-buff tinge seen in many adult female Red-wingeds; some solitary female Tricolored Blackbirds may not be identifiable with certainty. Immature males are similar to a winter adult male but the gray-buff feather edges are more obvious.

female

male

FACT FILE

LENGTH 8.75 in (22 cm)

FOOD Mainly seeds, with invertebrates in spring and summer

HABITAT Farmland and wet grassland

STATUS Locally common resident

VOICE Song is a series of croaking, screechy phrases. Call is a soft *chuk*

The Tricolored Blackbird is present year-round in California, where the vast majority of the population resides. It forms flocks outside the breeding season in particular.

male
life-size

Eastern Meadowlark

Sturnella magna

The Eastern Meadowlark is easily identified in eastern North America as no similar species occur here. However, where its range overlaps in the middle of the continent with the Western Meadowlark (p.302), separating the two is hard. Regional plumage variation adds to the problem. In any given location, Eastern Meadowlark sexes are similar. Adults have marbled brown upperparts and wings. The head pattern comprises buff cheeks, a dark crown, and a dark stripe behind the eye. The pale supercilium is yellow in front of the eye and white behind it, and the yellow throat is bordered by an entirely white malar stripe (the base of this is yellow in the Western Meadowlark) and a black chest band. The yellow-tinged underparts have dark spots on the flanks, and grade to white on the belly. In winter adults and juveniles, the black chest band is masked by pale feather margins. In flight, all birds show more white in the outer tail than does the Western Meadowlark.

The Eastern Meadowlark is present year-round in southern and southeastern North America; northern populations are present in their breeding range mainly from April to September, and migrate south in fall. Males often sing from roadside posts.

adult

FACT FILE

LENGTH 9.5 in (24 cm)

FOOD Mainly seeds, with invertebrates in spring and summer

HABITAT Grassland habitats

STATUS Widespread and common, resident in the south but a summer visitor in the north

VOICE Song starts with a few whistles, followed by a fluty *tswee-tswee-tsweuu*. Call is a buzzing *drzzt*

Western Meadowlark

Sturnella neglecta

The Western Meadowlark is the western counterpart of the Eastern Meadowlark (p.300) and has similar habits and habitat preferences. Regional variation in plumage exists but in any given location the sexes are similar. Adults have marbled brown upperparts and wings. The head pattern comprises buff cheeks, a dark crown, and a dark stripe behind the eye. The pale supercilium is yellow in front of the eye and white behind it, and the throat and much of the malar stripe are yellow (the malar stripe is white in the Eastern Meadowlark); the throat is defined below by a black chest band. The yellow-tinged underparts have dark spots on the flanks, and grade to white on the belly. In winter adults and juveniles, the black chest band is masked by pale feather margins. In flight, all birds show less white in the outer tail than does Eastern Meadowlark.

The Western Meadowlark is present year-round in the center of its range but present only as a breeding species farther north, mainly from March to September. Northern breeders move south in fall and the winter range extends to Central America.

adult
life-size

FACT FILE

LENGTH 9.5 in (24 cm)

FOOD Mainly seeds, with invertebrates in spring and summer

HABITAT Grassland habitats

STATUS Widespread and common; resident across much of its range but a summer visitor in the north, and a winter visitor farther south

VOICE Song comprises a few whistles and a fluty *tsoo-weet-tseleuu*. Call is a dry *tchuk* or a rattle

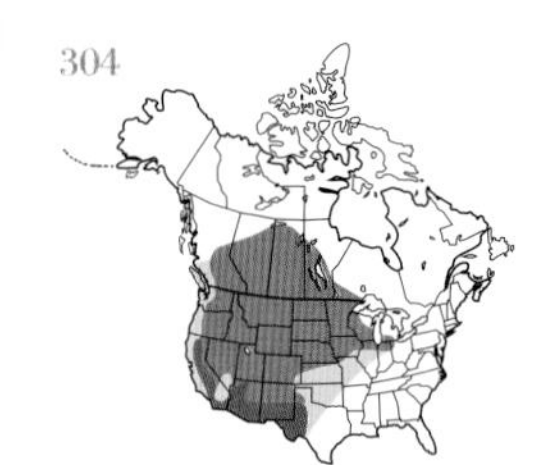

Yellow-headed Blackbird

Xanthocephalus xanthocephalus

female

The Yellow-headed Blackbird is a distinctive songbird. Males are subtly larger than females, and their plumages differ. Adult males have a yellow hood and breast, emphasizing the black patch through the eye. The plumage is otherwise black except for a striking white wing patch. Adult females have a yellowish-buff face, supercilium and breast, and otherwise unstreaked gray-brown plumage. Juveniles have a yellow-buff hood and breast, two indistinct pale wingbars, and otherwise gray-brown plumage with subtle yellow feather edges. By their first winter, immature females are similar to an adult female; and immature males are similar to an adult female but with a yellower head and a hint of a white wing patch.

male

FACT FILE

LENGTH 9–9.5 in (23–24 cm)

FOOD Mainly seeds, with invertebrates in spring and summer

HABITAT Wetland grassland and marshes

STATUS Widespread and common summer visitor; local in winter

VOICE Song is a varied series of harsh, screeching chatters. Call is a musical *kdek*

The Yellow-headed Blackbird is a summer breeding species across much of its range, present mainly from May to August. It spends the rest of the year usually in Mexico, although small numbers do winter in southern U.S.A. Outside the breeding season it forms flocks.

adult
life-size

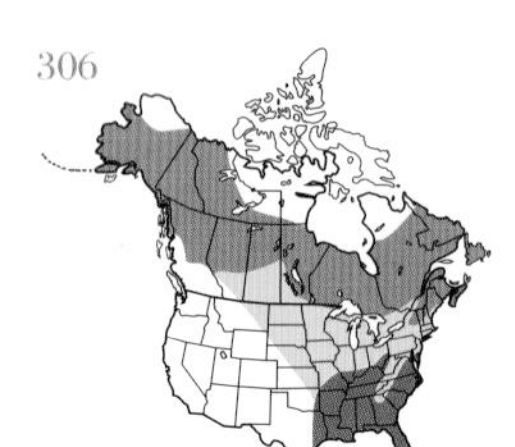

Rusty Blackbird

Euphagus carolinus

The Rusty Blackbird is a pale-eyed, slender-billed songbird. The sexes are dissimilar. Summer adult males are uniformly blackish, with a green sheen seen in good light. In fall, adult males have rusty brown feather margins over much of the body; these wear away by late winter to reveal pristine black feathers. Summer adult females are dark gray-brown overall, darkest on the cap, wings, and tail. In fall, adult females have rusty-brown edges to many feathers, especially on the head and back; these wear away in time. Immatures are similar to their respective winter adults.

female

The Rusty Blackbird is present in its northern breeding range mainly from May to September. It spends the rest of the year in southeast U.S.A. The species is probably easiest to see in the winter months.

breeding male

1st-winter male life-size

FACT FILE

LENGTH 9 in (23 cm)

FOOD Mainly seeds, with invertebrates in spring and summer

HABITAT Northern boggy forests in summer; open woodland and wetland margins in winter

STATUS Fairly common, but numbers have declined catastrophically in recent decades

VOICE Song includes a series of chuckling, chattering phrases, and ends with a whistle. Call is a soft *tchuk*

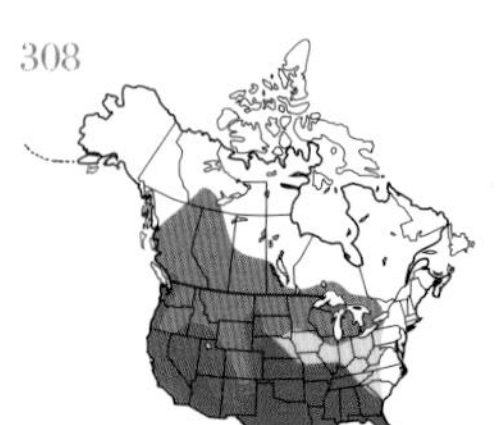

Brewer's Blackbird

Euphagus cyanocephalus

Brewer's Blackbird is a familiar suburban bird in the west of the continent. The sexes are dissimilar. Summer adult males have black plumage but, in good light, the head has a purple sheen and the back, wings, and breast have a greenish-blue sheen. The bill is slender and the eye has a pale iris. Winter adult males have brownish margins to their black feathers; these wear off by early spring. Adult females have dark gray-brown body plumage and a subtly darker tail and wings. In most individuals, the eye has a dark iris. Immatures are similar to their respective winter adults.

male life-size

FACT FILE

LENGTH 9 in (23 cm)

FOOD Invertebrates, seeds, and berries

HABITAT Open country, farmland, and parks

STATUS Widespread and common. Resident in the center of its range; summer and winter visitor elsewhere

VOICE Song comprises harsh, squeaky whistles. Call comprises *tchuk* notes

Brewer's Blackbird is a year-round resident in the middle of its extensive western range. In the breeding season, the range expands north and east; in the fall, these birds migrate south, and the winter range covers southern and southeastern U.S.A. and Central America. The species range is, to a degree, linked to environments altered by man, namely farmland and suburban developments.

Common Grackle

Quiscalus quiscula

Although superficially similar to a blackbird, the Common Grackle has a longer bill and a graduated tail that is folded lengthways and keeled. Males are larger than females and also differ in their plumage. Adult males across much of the range are black overall, with a blue sheen to the head, neck, and chest, a bronzed sheen on the body, and a bluish-purple tinge to the wings and tail. In birds from the southeast, the head and body have a purplish sheen. In all males, the eye has a pale iris. Adult females are similar to an adult male but duller, and they lack an obvious sheen. Juveniles are similar to an adult female but browner overall, and darkest and dullest on the wings and tail; the eye has a dark iris.

male

female

The Common Grackle is present year-round in southeast U.S.A., but in summer (mainly from May to September) its range extends north and east. It is easy to see in suburban environments and on farmland, and in spring males perform an interesting display.

male
life-size

FACT FILE

LENGTH 12.5 in (32 cm)

FOOD Invertebrates, seeds, and berries

HABITAT Open woodland, farmland, parks, and gardens

STATUS Widespread and common. Summer visitor in the north of its range; present year-round in the south

VOICE Song comprises harsh, grating phrases. Call is a sharp *tchuk*

Boat-tailed Grackle

Quiscalus major

The Boat-tailed Grackle is a large, slender-bodied songbird, the males of which have a very long paddle-shaped tail. The sexes are dissimilar in plumage terms and males are larger than females. Adult males are black overall, with a bluish sheen on the head and a greenish sheen on the body. Adult females have reddish-brown body plumage with a subtly darker tail and wings; the tail is relatively shorter than that of the male. In adults of both sexes, the iris is pale in Atlantic coast birds but browner in Florida and Gulf Coast birds. Juveniles are similar to an adult female but duller, and with a dark iris.

The Boat-tailed Grackle is present year-round in a narrow coastal belt along the east and southeast coasts of North America. It is generally bold and easy to observe.

male
life-size

FACT FILE

LENGTH 15–17 in (38–43 cm)

FOOD Invertebrates, seeds, and berries

HABITAT Coastal wetlands

STATUS Locally common resident with a restricted range

VOICE Song is a mix of peculiar dry rattles, hisses, and chattering notes. Call is a soft *tchek*

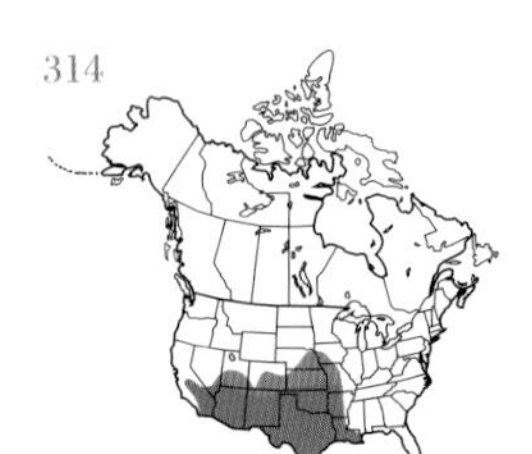

Great-tailed Grackle

Quiscalus mexicanus

The Great-tailed Grackle is the widespread western and inland counterpart of the Boat-tailed Grackle (p.312). All birds have a dagger-like bill; males are larger than females and have a long, graduated tail. Adult males are black overall with a bluish-violet sheen on the body. Adult females have brown body plumage, palest on the throat, chest, and supercilium, and a subtly darker tail and wings. Adults of both sexes have a pale iris. Juveniles are similar to an adult female but with streaked underparts and a dark iris.

The Great-tailed Grackle was once primarily a Mexican specialty but in recent decades its range has expanded dramatically. Now it is present year-round in the southwest U.S.A., and its range expands farther north in spring, with breeding birds present there mainly from April to September. Outside the breeding season it forms flocks. Being bold and noisy, it is an easy species to observe.

male

FACT FILE

LENGTH 15–18 in (38–45.5 cm)

FOOD Invertebrates, seeds, and berries

HABITAT Open habitats, including farmland, parks, and gardens

STATUS Locally common resident and partial migrant

VOICE Song is a mix of mechanical-sounding notes, including rattles, whistles, and noises resembling crackling electrical static. Call is a soft *tchut*

male
life-size
female

Bronzed Cowbird

Molothrus aeneus

The Bronzed Cowbird is a plump-bodied dark songbird. The sexes are dissimilar. Adult males have black plumage with a bronze sheen on the hood and back, and a blue sheen to the wings and tail. Adult females have dark brown plumage that is darkest on the wings and tail; the plumage lacks an obvious sheen. All adult birds have red eyes. Juveniles are similar to an adult female but the eye is dark.

With a mainly Mexican range, the Bronzed Cowbird is present as a breeding species from southern Texas to southern California mainly from May to August. It spends the rest of the year in Central America, where the species is resident. It is a nest parasite of other songbirds, undertaking no parental care of its offspring.

male

male

FACT FILE

LENGTH 8.75 in (22 cm)

FOOD Invertebrates, seeds, and berries

HABITAT Farmland and open country

STATUS Local summer visitor

VOICE Song is a series of strange-sounding gurgles and squeaks. Call is a harsh *tchuk*

female life-size

Brown-headed Cowbird

Molothrus ater

The Brown-headed Cowbird is a familiar plump-bodied songbird. The sexes are dissimilar. Adult males have a dark brown hood and otherwise blackish plumage with a green sheen. Adult females have dull gray-brown body plumage with a pale throat, malar stripe, eye surround, and supercilium; the wings and tail are subtly darker than the body. Juveniles are similar to an adult female, but the underparts are streaked, the back feathers have pale margins, and the wings show indistinct pale wingbars.

The Brown-headed Cowbird is present year-round in the south of its range. Elsewhere it is a migrant summer breeder across much of northern North America, present mainly from April to August. It is a nest parasite of other songbirds and does not undertake parental care of its offspring.

FACT FILE

LENGTH 7.5 in (19 cm)

FOOD Invertebrates, seeds, and berries

HABITAT Farmland and open country

STATUS Widespread and common summer visitor across northern North America; resident in the south

VOICE Song comprises a few strangled gurgles followed by a thin whistle. Call is a rattling *kerrk*

female
life-size

Orchard Oriole

Icterus spurius

The Orchard Oriole is a colorful, slim-bodied songbird with a slender downcurved bill. The sexes are dissimilar. Adult males have a black hood, chest, back, and tail, and a reddish-chestnut rump, underparts, and "shoulders." The black wings have a white wingbar and white edges to the flight feathers. All females are yellow overall but with a dull olive-yellow back and tail; the wings are subtly darker than the body, with two white wingbars and white edges to the flight feathers. Immature males are similar to a female but have a black face and throat.

female

1st-spring male

The Orchard Oriole is present as a breeding species in eastern North America, mainly from May to August. It spends the rest of the year in Central America. As orioles go, it is a fairly bold species and is usually easy to observe.

male life-size

FACT FILE

LENGTH 7.25 in (18.5 cm)

FOOD Invertebrates, seeds, and berries

HABITAT Open woodland and wooded parks

STATUS Widespread and locally common summer visitor

VOICE Song is a jaunty series of fluty whistles; call is a harsh chatter

Hooded Oriole

Icterus cucullatus

The Hooded Oriole is a long-tailed, slim-bodied songbird with a slender downcurved bill. The sexes are dissimilar. Adult males have orange body plumage with a black face, throat, back, and tail. The black wings have two white wingbars and white edges to the flight feathers. All females have an olive-yellow tail and upperparts, and a dull yellow face and underparts, palest on the flanks. The dark wings have two white wingbars and white edges to the flight feathers. Immature males in fall are similar to an adult female but by their first spring they have acquired an adult male's black face and throat.

The Hooded Oriole is present as a breeding species in its southwestern range mainly from April to August. It spends the rest of the year in Mexico. Partly because it favors open habitats, it is usually easy to observe.

FACT FILE

LENGTH 8 in (20 cm)

FOOD Invertebrates, seeds, and berries

HABITAT Open woodland, often near water

STATUS Locally common summer visitor

VOICE Song is a short series of harsh, warbling phrases. Call is a harsh *tchuut*

Bullock's Oriole

Icterus bullockii

Bullock's Oriole is a colorful and well-marked songbird. The sexes are dissimilar. Adult males have an orange face, supercilium, and underparts, with a black crown, eye stripe, nape, and back, and a narrow black line on the throat. The black wings show a broad white panel and white edges to the flight feathers. The rump is orange and the tail is black with orange outer feathers. All females have an olive-gray back; a yellow hood, breast, and flanks; and otherwise mostly whitish underparts. The dark wings have two white wingbars and white edges to the flight feathers. The rump and tail are yellowish. Immature males in fall are similar to an adult female but by their first spring they have acquired the adult male's black mask and throat.

Bullock's Oriole is present as a breeding species in western North America mainly from May to August. It spends the rest of the year in Mexico. Despite its colorful appearance, it can be hard to spot as it forages for insects in dappled foliage.

female

FACT FILE

LENGTH 8.75 in (22 cm)

FOOD Invertebrates, seeds, and berries

HABITAT Open woodland, usually near water

STATUS Widespread and common summer visitor

VOICE Song is a series of shrill whistles, followed by more tuneful *tuwee-weep* notes. Call is a dry *tchup*

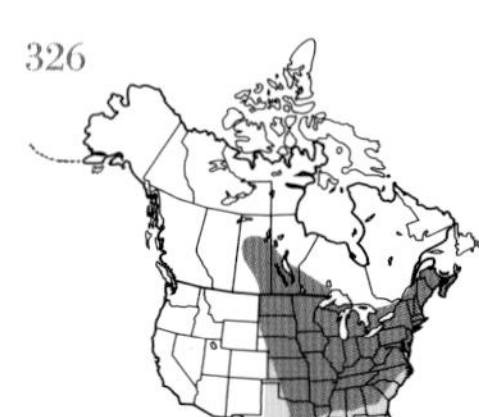

Baltimore Oriole

Icterus galbula

The Baltimore Oriole is a colorful, slim-bodied songbird. The sexes are dissimilar. Adult males have a black hood and back, and orange underparts and "shoulders." The black wings show a white wingbar and white edges to the flight feathers. The rump is orange, and the tail is orange with a dark base and central feathers. Adult females recall an adult male but are less colorful overall, and black elements of the plumage on the hood and back are replaced by variably dark olive-brown. The "shoulder" patch is replaced by a second white wingbar, and the rump and tail are buffish orange. Immatures are similar to an adult female; immature males are a richer orange than immature females (which are yellower) and the upper white wingbar is more pronounced.

female

FACT FILE

LENGTH 8.75 in (22 cm)

FOOD Invertebrates, seeds, and berries

HABITAT Open woodlands and wooded parks

STATUS Widespread and common summer visitor; local and scarce in winter

VOICE Song is a series of tuneful, whistled *chewdi* phrases. Call is a rattle

The Baltimore Oriole is present as a breeding species in the eastern half of North America mainly from May to August. It spends the rest of the year mainly in Central or South America, with very small numbers wintering in southeast U.S.A. The species is usually easy to observe.

female

male life-size

Scott's Oriole

Icterus parisorum

Scott's Oriole is a slim-bodied songbird with a slender, pointed bill. The sexes are dissimilar. Adult males have a black back, hood, and breast, and otherwise yellow underparts. The wings are black overall but with a yellow "shoulder," a white wingbar, and white edges to the flight feathers. From above, the tail is black with yellow sides to the basal half of the outer feathers; from below, the tail is yellow. Adult females recall an adult male but black elements of the plumage on the head and back are mottled olive-gray. On the wing, the "shoulder" patch is replaced by a second wingbar. Immatures are similar to their respective adults but duller and less strikingly marked.

male

1st-spring male

Scott's Oriole is present as a breeding species in southwestern desert habitats mainly from April to August; it favors areas where yuccas (*Yucca* spp.) and junipers (*Juniperus* spp.) flourish. It spends the rest of the year in Mexico.

FACT FILE

LENGTH 9 in (23 cm)

FOOD Invertebrates, seeds, and berries

HABITAT Open desert scrub

STATUS Locally common summer visitor

VOICE Song comprises short bursts of warbling, fluty whistles. Call is a harsh *tchek*

female
life-size

Gray-crowned Rosy-Finch

Leucosticte tephrocotis

The Gray-crowned Rosy-Finch is a hardy songbird of challenging environments. The sexes are dissimilar and the plumage shows subtle regional variation. All adult males have a black forecrown and throat, and rosy-pink wing coverts, flight feather margins, rump, and belly. In birds that breed in the interior, the rear of the crown is gray and the cheeks and breast are brown. Coastal breeders are similar, but the cheeks as well as the rear of the crown are gray. Birds that breed on Bering Sea islands are similar to coastal birds but the back and breast are blackish brown. Adult females are similar to their respective regional males but paler and much less colorful. In all adults the bill is black in summer but yellow in winter. Juveniles are brown overall with pale edges to the wing feathers and a dark forecrown.

The Gray-crowned Rosy-Finch is a breeding visitor in the north of its range mainly from May to August; these birds move south in fall. Birds from more southerly mountain ranges and Bering Sea birds are year-round residents. It forms flocks in winter.

FACT FILE

LENGTH 6 in (15 cm)

FOOD Invertebrates, seeds, and berries

HABITAT Arctic tundra and short alpine vegetation

STATUS Locally common, with different summer and winter ranges

VOICE Song is a descending series of whistled *chew* notes. Call is a harsh *chew*

Black Rosy-Finch

Leucosticte atrata

female

The Black Rosy-Finch is a smart-looking, plump songbird. The sexes are dissimilar. Adult males have mostly blackish plumage with a rosy-pink belly and patch on the wings, and a gray rear crown. Adult females are similar, but with muted colors and black elements of the male's plumage tinged gray. Juveniles are gray overall, with two pink wingbars and pale edges to the flight feathers.

The Black Rosy-Finch is present in its Rocky Mountain range year-round. However, there is considerable altitudinal migration and dispersal during the winter months as a result of unpredictable snowfall. Outside the breeding season it forms flocks that feed at the edges of snow patches.

FACT FILE

LENGTH 6 in (15 cm)

FOOD Invertebrates, seeds, and berries

HABITAT Alpine habitats, above the treeline

STATUS Locally common

VOICE Song is a series of *chuup* notes. Call is a sharp *chuup*

male
life-size

Brown-capped Rosy-Finch

Leucosticte australis

The Brown-capped Rosy-Finch is a hardy specialty of high-altitude mountain habitats. The sexes are dissimilar. Adult males have largely cinnamon-brown body plumage with rosy-pink wing coverts and wing feather margins, a pink rump and belly, and a dark cap. Adult females are more uniformly brown than an adult male, with very little pink in their plumage. Juveniles are similar to an adult female but much grayer overall.

The Brown-capped Rosy-Finch is present year-round in its eastern Rocky Mountain range. Outside the breeding season it forms flocks; harsh winter weather causes some altitudinal migration and dispersal. The species often feeds at the edge of melting snow.

female

male life-size

FACT FILE

LENGTH 6 in (15 cm)

FOOD Invertebrates, seeds, and berries

HABITAT Alpine habitats, above the treeline

STATUS Locally fairly common

VOICE Song comprises a series of harsh *chew* notes. Call is a harsh *chew*

House Finch

Haemorhous mexicanus

The House Finch is a familiar little songbird. The sexes are dissimilar. In adult males the forehead, broad supercilium, breast, and rump are bright red. The back, nape, and center of the crown are brown, and the dark brown wings show two white wingbars and pale edges to the flight feathers. The belly and undertail are white with bold streaking on the flanks. Adult females and immatures are gray-brown overall and streaked; the subtly darker wings show two pale wingbars and pale edges to the flight feathers.

female

Once restricted to the west of the continent, the House Finch is now present year-round across much of central and southern North America. It lives up to its name and is often associated with gardens, where it visits birdfeeders.

FACT FILE

LENGTH 6 in (15 cm)

FOOD Invertebrates, seeds, and berries

HABITAT Wide range of wooded habitats, including parks and gardens

STATUS Widespread and common resident

VOICE Song is a series of twittering, chattering phrases ending with a harsh *wheert*. Call is a shrill *whee-ert*

male life-size

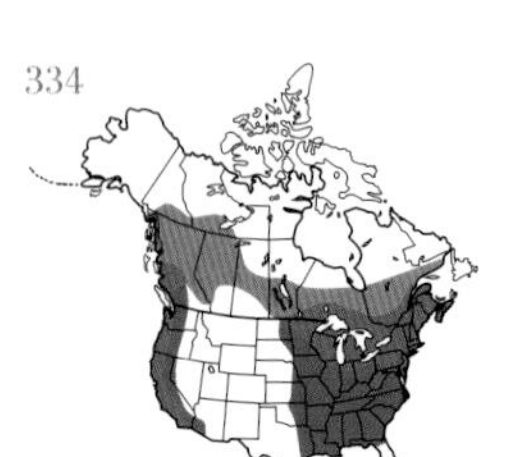

Purple Finch

Haemorhous purpureus

female

The Purple Finch is a small, plump-bodied songbird with a conical bill and proportionately large head. The sexes are dissimilar. Adult males are reddish pink on the head and breast, grading to streaked reddish brown on the back. The subtly darker wings have two rather indistinct pinkish-buff wingbars and buff margins to the flight feathers. The flanks are flushed pink and streaked. The belly and undertail are white in birds from the east of the range but gray in western birds. Adult females and juveniles have streaked gray-brown upperparts, and pale underparts that are dark-streaked except on the undertail. The brown head has a pale supercilium, submustachial stripe, and throat; the subtly darker ear coverts and malar stripe are more pronounced in eastern birds than in western ones.

The Purple Finch is a breeding visitor to the north of its range, mainly from May to August; in winter these birds spread across much of eastern U.S.A. In the west and northeast of its range it is present year-round.

FACT FILE

LENGTH 6 in (15 cm)

FOOD Invertebrates, seeds, and berries

HABITAT Conifer and mixed forests

STATUS Widespread and common summer breeder, winter visitor, and resident

VOICE Song is a series of warbling phrases. Call is a sharp *pik*

male life-size

Cassin's Finch

Haemorhous cassinii

In many ways, Cassin's Finch is the montane counterpart of the Purple Finch (p.334). The sexes are dissimilar. Adult males have a reddish-pink crown, throat, breast, and rump. The nape and back are streaked brown, and the subtly darker wings show two pinkish wingbars and pale edges to the flight feathers. The otherwise pale underparts are flushed pink on the flanks and streaked on the flanks and undertail (these are unstreaked in Purple Finch). Adult females and juveniles have streaked brown upperparts, and subtly darker wings showing two pale wingbars. The pale underparts, including the undertail, are marked with fine dark streaks.

Cassin's Finch is a conifer specialist that is present year-round in much of its western montane range. During the breeding season its range extends farther north and to higher altitudes. Outside the breeding season it forms flocks; harsh winter weather often forces birds to move to lower elevations and latitudes.

female

FACT FILE

LENGTH 6.25 in (16 cm)

FOOD Invertebrates, seeds, and berries

HABITAT Upland conifer forests

STATUS Locally common resident and partial migrant

VOICE Song is a series of warbling whistles. Calls include a soft *tch-wu*

male life-size

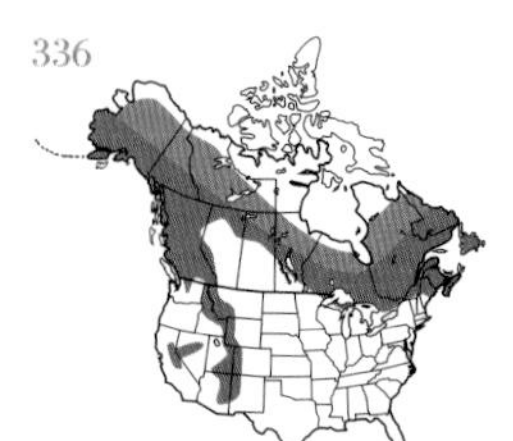

Pine Grosbeak

Pinicola enucleator

The Pine Grosbeak is a plump, stubby-billed songbird. The sexes are dissimilar. Adult males are pinkish red overall, variably flushed gray on the flanks and belly. The tail is dark and the dark wings show two white wingbars. Adult females show a suggestion of the male's plumage pattern, but are gray overall with a variably yellowish head, back, and rump. Juveniles are brown overall with two pale wingbars; by their first year they are similar to an adult female.

FACT FILE

LENGTH 9 in (23 cm)

FOOD Mainly shoots, buds, and seeds

HABITAT Conifer forests

STATUS Widespread and fairly common resident and partial migrant

VOICE Song is a series of whistling *pwee* phrases. Call is a whistled *piew*

male

The Pine Grosbeak is a northern specialist that is present year-round throughout much of northern North America and in western mountains. In the breeding season its range expands north; in winter, birds are often forced to move south. Outside the breeding season it is sometimes seen in small flocks.

female life-size

male

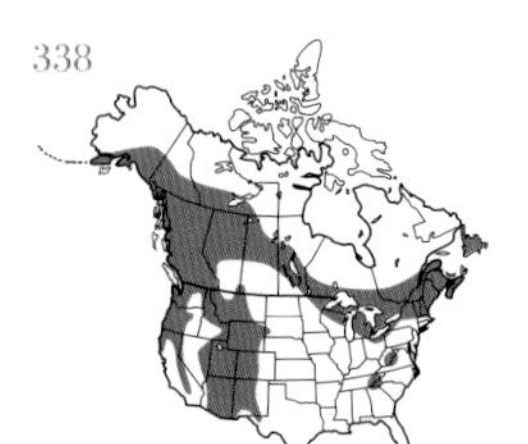

Red Crossbill

Loxia curvirostra

female

The Red Crossbill has a bill with cross-tipped mandibles that are used to extract seeds from conifer cones; this distinctive feature is shared only with the White-winged Crossbill (p.339). The sexes are dissimilar. Adult males have red plumage that is darkest and brownest on the wings and tail. Confusingly, the occasional individual has subtle pale wingbars, leading to potential for confusion with the White-winged. Adult females have yellow-green plumage that is darkest and brownest on the wings and tail. Juveniles are streaked brown, and darker above than below. By the first year, immature males are similar to an adult female but the plumage is tinged orange-yellow; first-year females are similar to an adult female.

The Red Crossbill is present year-round in northern and western conifer forests.

male
life-size

FACT FILE

LENGTH 6.25 in (16 cm)

FOOD Conifer seeds

HABITAT Conifer forests

STATUS Locally common resident

VOICE Song often includes shrill *kip* notes, and usually ends in a buzzing trill. Call is a sharp *kip-kip*, often given in flight

White-winged Crossbill

Loxia leucoptera

female

The White-winged Crossbill is a striking finch whose bill has cross-tipped mandibles. The bold white wingbars are much more striking than in variant Red Crossbills (p.338) and allow identification with certainty. The sexes are dissimilar. Adult males have bright pinkish-red body plumage (much paler and pinker than in a male Red Crossbill) that is palest on the belly and flanks. The dark wings have two striking white wingbars, and the tail is blackish. Adult females have streaked olive-yellow body plumage and dark wings with a wingbar pattern like that of a male. Juveniles are streaked gray-brown and darker above than below; the dark wings show two wingbars but these are less distinct than on an adult. By the first year, an immature male recalls an adult male but pink elements of the plumage are tinged yellow; immature females are similar to an adult female.

Most White-winged Crossbills are resident year-round in northern larch and spruce forests They form roving flocks outside the breeding season.

FACT FILE

LENGTH 6.5 in (16.5 cm)

FOOD Conifer seeds

HABITAT Conifer forests

STATUS Widespread and locally common resident

VOICE Song is a series of trills and whistles. Call is a sharp *chip-chip*, usually given in flight

male life-size

Common Redpoll

Acanthis flammea

The Common Redpoll is a compact little finch. The sexes are dissimilar. Adult males have streaked gray-brown upperparts, and subtly darker wings showing two white wingbars and white margins to the flight feathers. The streaked gray-brown head has a red forecrown and black chin. The breast and flanks are suffused with pinkish red, which is more intense in summer than winter. The underparts are otherwise whitish but streaked on the flanks. Adult females are similar to an adult male but lack the red flush on the breast at all times. Juveniles are streaked and buffish brown, and lack the adult's red forecrown; this feature is acquired by early winter.

The Common Redpoll is present year-round in boreal forests across the region. Birds that breed farther north in the Arctic are present there mainly May to August; outside the breeding season they form nomadic flocks that move south, roaming in search of reliable sources of alder and birch seeds.

FACT FILE

LENGTH 5.25 in (13.5 cm)

FOOD Invertebrates in summer; seeds and buds at other times

HABITAT Northern forests

STATUS Widespread and common resident, summer breeder, and winter visitor

VOICE Song is a series of trills and twitters. Call is a dry rattle

female

male life-size

Hoary Redpoll

Acanthis hornemanni

The Hoary Redpoll is a paler, dumpier version of a Common Redpoll (p.340), with a stubby little bill and an unstreaked white rump. The sexes are dissimilar. Adult males have a dark-streaked buffish-white back and nape, with streaking on the back, nape, and rear of the crown. The head has a red forecrown, a pale gray face, and a small black chin. The underparts are white with faint dark streaks on the flanks; the breast is suffused very pale pink in spring and early summer. Adult females are similar to a winter adult male but with slightly bolder streaking on the flanks. Juveniles are similar to an adult female but more heavily streaked on the flanks and suffused buff on the face.

female

The Hoary Redpoll is such a hardy species that some birds remain in the high Arctic throughout the year. However, outside the breeding season many form roaming flocks that move south in search of food, sometimes mixing with Common Redpolls.

FACT FILE

LENGTH 5.5 in (14 cm)

FOOD Invertebrates in summer; seeds and buds at other times

HABITAT Arctic tundra scrub in the breeding season; northern forests in winter

STATUS Widespread and locally common

VOICE Song is a series of trills and twitters. Call is a dry rattle. Both are similar to those of the Common Redpoll (p.340)

male life-size

Pine Siskin

Spinus pinus

female

The Pine Siskin is a compact little finch with a pointed conical bill. The sexes are dissimilar. Adult males have streaked gray-brown upperparts and dark-streaked whitish underparts. Some birds have a yellow wash to the underparts and rump. The dark wings have two wingbars, the upper one white and the lower one tinged yellow. The yellow edges to the flight feathers appear as an obvious bar in flight. Adult females are similar to a dull adult male but any yellow coloration is much less intense. Juveniles are similar to an adult female but are suffused buffish overall.

The Pine Siskin is present year-round in northern forests and in montane forests in the west. The summer breeding range extends farther north. Outside the breeding season roaming flocks move south and may turn up almost anywhere in the continent except the far southeast. The seeds of birch, alder, and spruce trees are particularly important as food in the winter months.

FACT FILE

LENGTH 5 in (12.5 cm)

FOOD Invertebrates in summer; seeds and buds at other times

HABITAT Conifer forests in spring and summer; conifer and mixed forests at other times

STATUS Widespread and common resident, summer breeder, and winter visitor

VOICE Song is a series of whistles and trills. Call is a buzzing *zhreee*

male life-size

Lesser Goldfinch

Spinus psaltria

The Lesser Goldfinch is a specialty of the southwest.

The sexes are dissimilar. Adult males have bright yellow underparts. Eastern birds have black upperparts, the wings showing a white wingbar and white at the base of the primaries and on the tertials; the white coloration on the outer feathers of the otherwise black tail is most obvious in flight. Western populations are similar but the back and nape are olive-green. Adult females have olive-green upperparts, with dark wings showing two indistinct wingbars, a white wing patch, and white edges to the tertials. Juveniles are similar to an adult female; by fall, immature males acquire a hint of an adult male's black cap and forehead.

The Lesser Goldfinch is a mainly Central American species that is present year-round in the south of its North American range. In the breeding season (mainly from April to August) its range extends farther north. Outside the breeding season it forms flocks that often feed on thistles.

FACT FILE

LENGTH 4.5 in (11.5 cm)

FOOD Invertebrates in summer; mainly seeds at other times

HABITAT Open dry woodland

STATUS Common resident and summer breeder

VOICE Song is a jumbled series of whistles, chirps, and squeaks. Call is a sharp, whistled *tee-oo*

Lawrence's Goldfinch

Spinus lawrencei

Lawrence's Goldfinch is a well-marked little finch. The sexes are dissimilar. Adult males have a black face, throat, and forecrown, and pale gray cheeks; the nape and back are gray in summer but olive-buff in winter. The dark wings have two yellow wingbars and yellow margins to the flight feathers. The breast is yellow, grading to gray on the rest of the underparts. A yellow rump is obvious in flight. Adult females recall an adult male in winter, but they lack any black on the face, are much paler overall, and have browner upperparts, especially in winter. Juveniles recall a heavily streaked winter female.

Lawrence's Goldfinch is a specialty of California, its range extending to northern Baja California. Although it is present year-round within its range, its precise occurrence is hard to predict as birds wander in search of reliable food supplies; thistle seeds are a favorite.

female

male
life-size

FACT FILE

LENGTH 4.75 in (12 cm)

FOOD Invertebrates in summer; mainly seeds other times

HABITAT Scrub-covered slopes

STATUS Local

VOICE Song is a series of whistling and tinkling notes, sometimes including mimicry. Call has a bell-like quality

American Goldfinch

Spinus tristis

winter male

The American Goldfinch is a distinctive songbird, males of which are very colorful. The sexes are dissimilar. Summer adult males have mainly yellow plumage with a black cap and forehead, a black tail, and black wings with a narrow yellow wingbar. The rump and undertail are white. Winter adult males recall a summer male but yellow elements of the plumage are yellow-buff, grading to grayish white on the belly. The dark wings show a pale patch on the greater coverts and a white wingbar. Adult females are similar to a winter male, but are brighter yellow overall in summer and grayer buff overall in winter. Juveniles are similar to a winter female.

The American Goldfinch is present year-round across much of the continent. In the breeding season its range extends farther north, and in winter it extends south throughout southern U.S.A. and into Mexico. Outside the breeding season the species forms flocks that are often seen feeding on thistle seedheads.

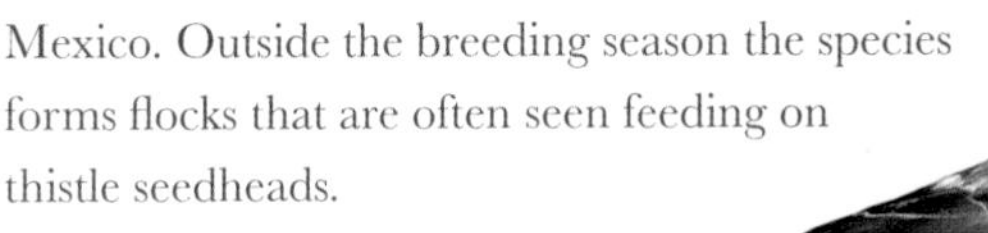

summer female

FACT FILE

LENGTH 5 in (12.5 cm)

FOOD Invertebrates in summer; mainly seeds at other times

HABITAT Open habitats, including weedy fields and scrub

STATUS Widespread and common resident and summer breeder

VOICE Song is a series of chattering notes and whistles. Calls include various whistling notes

summer male
life-size

Evening Grosbeak

Coccothraustes vespertinus

The Evening Grosbeak is a plump finch whose huge bill gives it a distinctive silhouette. The sexes are dissimilar. Adult males have a yellow forehead and supercilium, and an otherwise brown crown, face, and nape that grades to yellow on the mantle. The brown neck grades to golden yellow on the rest of the underparts. The tail is black and the wings are black with a white panel. Adult females are patterned like an adult male but the body plumage is mainly gray-buff, suffused yellow on the underparts. The tail is black with a white tip and the wings are black with two white patches. Juveniles are similar to their respective adults in terms of plumage pattern; juvenile females are similar in terms of color as well, but juvenile males are buffish brown overall.

male

female

FACT FILE

LENGTH 8 in (20 cm)

FOOD Invertebrates, seeds, and berries

HABITAT Conifer and mixed forests and woodland

STATUS Widespread and common resident and partial migrant

VOICE Song is a seldom-heard series of subdued whistling notes. Call is a soft *pee-irp*

The Evening Grosbeak is generally present year-round in its resident range, but in most years there is some movement southwards of nonbreeding flocks. In addition, large irruptive movements related to food shortages occur periodically. The species often visits birdfeeders in the winter months.

male life-size

House Sparrow

Passer domesticus

female

The House Sparrow is a familiar urban songbird. The sexes are dissimilar. Adult males have mainly chestnut-brown upperparts, streaked on the back, and chestnut-brown wings with a white wingbar. The crown, cheeks, and rump are gray. The throat and breast are black, and the underparts are otherwise pale gray. In winter, pale feather margins (which wear away) render the coloration muted at first. Adult females have mainly gray-buff upperparts, including the crown, with dark streaking on the back; the head has a buff supercilium. The underparts are pale gray-buff. Juveniles are similar to an adult female but the markings are less distinct.

Introduced by European settlers, the House Sparrow is now present year-round across much of the continent. It is almost always found in association with suburban and urban parks and gardens, as well as on farmland, where it congregates in flocks around buildings. In many situations it becomes very tame.

FACT FILE

LENGTH 6 in (15 cm)

FOOD Mostly seeds, but also some invertebrates in spring and summer

HABITAT Farmland, town parks, and gardens

STATUS Widespread and common resident

VOICE Song comprises a jumbled series of chirping notes. Calls include various chirps and shrill notes

summer male
life-size

Eurasian Tree Sparrow

Passer montanus

The Eurasian Tree Sparrow is similar to a House Sparrow (p.348) in terms of its size and proportions, but has distinctive plumage. The sexes are similar. Adults have brown upperparts with dark streaking on the back; the wings have dark feather centers and show two white wingbars. The chestnut cap is offset by the bold black patch and bib on the otherwise whitish cheeks and side of the head. The underparts are otherwise grayish white. Juveniles are similar to an adult but the markings are duller and less distinct, especially on the face.

adult

The Eurasian Tree Sparrow was introduced from Europe in the 19th century and is now an established resident, mainly in west Illinois and neighboring states. It is semicolonial when nesting and is found in flocks outside the breeding season.

adult life-size

FACT FILE

LENGTH 6 in (15 cm)

FOOD Mostly seeds, but some invertebrates in spring and summer

HABITAT Farmland and open country with scrub, and suburban parks

STATUS Local resident

VOICE Song includes various whistles and chirps. Calls include chirps and a sharp *tik-tik* in flight

Index

Photographic credits

All photographs used in this book were taken by **Brian E. Small** of **Nature Photographers Ltd** with the exception of:

From Nature Photographers Ltd: **Klaus Bjerre**: Northern Shrike, 47 (top). **Andrew Cleave**: Evening Grosbeak, 346 (top). **David Osborn**: Clark's Nutcracker, 70 and 71 (bottom); Black-billed Magpie, 73; Common Raven, 78 (top). **Roger Tidman**: Bank Swallow, 87 (top).

From Alamy: **Glenn Bartley**/All Canada Photos/Alamy Stock Photo: Northern Shrike, 46; Yellow-bellied Flycatcher, 14 (left); Plumbeous Vireo, 52 (bottom); Cordilleran Flycatcher, 23 (bottom); Smith's Longspur, 177. **Rick & Nora Bowers**/Alamy Stock Photo: Rusty Blackbird 1st winter, 307; Brown-crested Flycatcher, 33. **Kitchin & Hurst**/All Canada Photos/Alamy Stock Photo: Kirtland's Warbler female, 202. **Cal Vornberger**/Alamy Stock Photo: Connecticut Warbler immature, 195; Saltmarsh Sparrow, 260 (top); Eastern Towhee female, 235.

From other sources: **Garth McElroy**: White-winged Crossbill male and female, 339.